Amar Bakdid
Bachir El Kihel

Thermal spray coatings and ultrasonic quality control

AF376634

Amar Bakdid
Bachir El Kihel

Thermal spray coatings and ultrasonic quality control

Thermal spraying to improve wear-resistant surface quality

ScienciaScripts

Imprint

Any brand names and product names mentioned in this book are subject to trademark, brand or patent protection and are trademarks or registered trademarks of their respective holders. The use of brand names, product names, common names, trade names, product descriptions etc. even without a particular marking in this work is in no way to be construed to mean that such names may be regarded as unrestricted in respect of trademark and brand protection legislation and could thus be used by anyone.

Cover image: www.ingimage.com

This book is a translation from the original published under ISBN 978-620-6-69752-7.

Publisher:
Sciencia Scripts
is a trademark of
Dodo Books Indian Ocean Ltd. and OmniScriptum S.R.L publishing group

120 High Road, East Finchley, London, N2 9ED, United Kingdom
Str. Armeneasca 28/1, office 1, Chisinau MD-2012, Republic of Moldova, Europe
Printed at: see last page
ISBN: 978-620-6-93244-4

Copyright © Amar Bakdid, Bachir El Kihel
Copyright © 2023 Dodo Books Indian Ocean Ltd. and OmniScriptum S.R.L publishing group

TABLE OF CONTENTS

TABLE OF CONTENTS 1

INTRODUCTION 3

Chapter I: Thermal projection 5

1. Tribological system. 6

2. Surface degradation (wear and corrosion). 6

3. Main wear patterns 7

    3.1. Wear due to abrasion, erosion and cavitation 8

    3.2. Fatigue and delamination wear 10

    3.3. Adhesive wear. 11

    3.4. Other types of wear. 12

4. The benefits of thermal spray surface coating. 13

5. Thermal spraying process. 13

    5.1. General principle. 13

    5.2. The different projection methods. 14

        5.2.1. Flame projection : 14

        5.2.1.1. "Powder flame 15

        5.2.1.2. "Flamme fil 15

        5.2.1.3. Hypersonic flame: High Velocity Oxyfuel Flame (HVOF). 16

        5.2.1.4. Detonation gun. 16

    5.3. Arc spraying between two wires : 17

    5.4. Plasma projection : 17

        5.4.1. Plasma. 17

        5.4.2. Principle of plasma spraying. 18

        5.4.3. The different plasma spraying processes. 18

        5.4.4. Some mechanical properties of plasma sprayed coatings. 19

    5.5. Applications. 19

    6. Conclusion. 19

1. Coating. 22

    1.1. Definition. 22

    1.2. Coating process. 22

    1.3. Mode of action. 22

1.4.   Main features. 23

1.5.   Advantages of the coating process. 23

2.   Coating adhesion. 23

2.1   Definition. 23

2.2   The challenges of measuring adhesion. 24

2.3   How to determine adhesion. 24

2.4   Mechanisms of adhesion of deposits to the substrate. 24

3.   Operating parameters for thermal spraying. 25

3.1   Relative speed torch - substrate. 27

3.2   Projection distance 29

3.3   Projection angle 30

3.4   No scanning 31

1.   Thermal gas spraying : 34

1.1.   Introduction. 34

1.2.   Hardware 34

1.2.1.   Spray gases and their storage 35
1.2.2.   Pressure reducing valves 36
1.2.3.   Gas hoses (flexible) 36
1.2.4.   Flashback arrestor 36
1.2.5.   Projection torches. 36
1.2.6.   Filler materials. 37

2.   Safety instructions. 38

3.   Application : 38

3.1.   Characterization methods : 40

3.2.   Results and discussion : 40

3.3.   Interpretation : 43

3.4.   Conclusion : 44

General conclusion : 46

Reference and Bibliography : 47

INTRODUCTION

Thermal spray coatings date back to 1909, with the invention by Schoop (who left the name "schoopage" to these techniques) relating to the projection of molten lead using a sprayer, followed by powdered lead through a flamme. Since 1914, the process has been used for industrial applications, mainly to protect against wear and corrosion, or for fins of manufacture. The first industrial applications appeared at the start of the First World War, to coat the rear face of shells with tin. The mechanical engineering industry began using this technique after the Second World War to refurbish worn parts. The need for more specific properties in many applications, particularly aeronautics, aerospace, the nuclear industry and mechanical engineering where components operate under severe conditions (internal stresses: mechanical stress, fatigue, fluage...; external stresses: friction, abrasion, temperature...; environmental stresses: corrosion, oxidation, chemical attack....), has led to the development of new thermal spraying techniques categorically classified into two families according to the heat source employed: flamme spraying, and it currently groups flamme-powder spraying, flamme-fil spraying, supersonic combustion spraying (HVOF: High Velocity Oxy-fuel Flame and HVAF: High Velocity Air-fuel Flame) and the detonation gun. The second family is electric arc spraying, and currently includes arc-fil spraying and soufflé arc plasma spraying.

Like most surface treatment processes, thermal spraying plays an important role in the mechanical engineering industries. It involves spraying fine particles onto a surface that has probably been prepared.

Thermal spray coatings are widely used to modify the physical, chemical and mechanical properties of industrial part surfaces, and to improve surface quality.

Today, therefore, the competitiveness of machines and products is no longer judged solely on their technical performance, but also on their reliability and longevity, as well as the cost of installation. It's an incomparable advantage to know what needs to be done to ensure that parts don't wear out, don't seize up, are more resistant to corrosion, and focus more on reducing the cost of improving the performance of processed parts.

The work we are going to present is as follows:

In the first part, we will review the literature on thermal spraying (Chapter 1). Next, we define the coating adhesion of a substrate and determine the operating parameters of thermal spraying (kinematic parameter) which influence the coating characteristics (Chapter 2).

The second part of the document will be completed by experimental tests we carried out in an industrial company using the method, and the rest of our test is completed in our laboratory at ENSAO (Plate Forme Technologique: P.F.T.2M) by ultrasonic testing to determine the importance of adhesion (chapter 3). We end our manuscript with a general conclusion in which we summarize the various results.

CHAPTER I: THERMAL SPRAYING

1. Tribological system.

A tribological system is defined as a mechanical system made up of two antagonistic materials in contact, animated by relative movements. These two solids evolve in an ambient medium and can be separated by an interposed film called the third body. In tribological terms, the third body [1] is an operator that transmits the load (or lift) of one body to the other, and accommodates most of the speed difference between two bodies by dissipative flow (friction). Third bodies partially or completely separate the first bodies. They are introduced into the contact either by kinematic entrainment (solid or liquid lubricants) or by in situ formation (loose particles) (Figure 1.1).

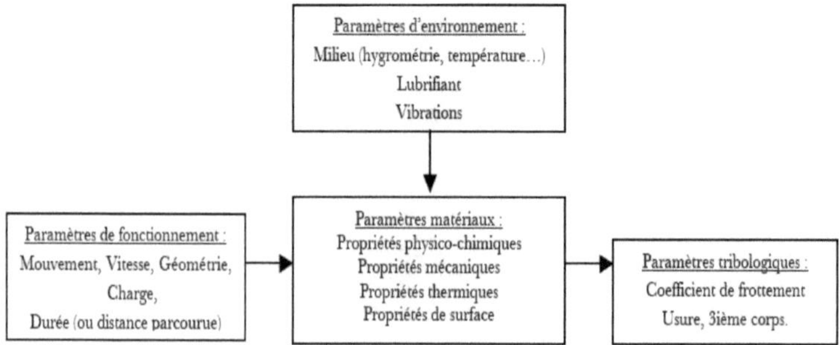

Figure 1.1: General diagram of a tribological system.

2. Surface degradation (wear and corrosion).

Wear is a complex set of phenomena, leading to the emission of debris with loss of mass, ribs and shape, and accompanied by physical and chemical transformations of surfaces.

It does not generally vary progressively with parameters such as speed, temperature or time. While some forms of wear are relatively regular, others are subject to very sharp jumps, sometimes in ratios of 1 to 100,000 or more, when certain critical values are exceeded.

Wear is generally resisted because of its negative effects, but it also has positive aspects. Sharpening a tool, finishing a surface by grinding, writing chalk on a blackboard or pencil on paper are all examples of useful abrasive wear.

Some forms of wear and tear are deliberately induced to combat others that would be far more devastating. When properly managed, the running-in of a mechanism, i.e. the operating phase when parts "learn to live together", causes wear that will prove "protective" in subsequent phases.

Most of the time, the overall wear of a mechanism is due to several processes acting simultaneously, more rarely to a well-defined and identifiable process. The effect of these simultaneous actions is often greater than the sum of the effects that would be produced by making the various processes act separately, and is sometimes referred to as "overadditivity" [2].

In the initial stages of friction, the surface screens are subjected to all the mechanisms inherent in friction (thermal, mechanical or chemical effects). Subsequently, these mechanisms take turns or interpenetrate depending on the multiple conditions present.

In general, a part undergoes 3 stages of wear throughout its life:

- Running-in: rapid wear with steadily decreasing wear rate.

- Normal operation: service life (low, constant wear).

-Ageing and death: increasing wear rate.

3.   Main wear modes

First of all, it is worth recalling the main wear patterns observed on metallic and non-metallic materials. The classification of wear patterns proposed in this paragraph is inspired by that of Stachowiak et al, 2001 [3].

It distinguishes several wear modes: abrasive, erosive, cavitational, adhesive, fatigue and delamination, and other more specific wear modes. It introduces the basic concepts used in tribology, as well as important quantities. It will also provide food for thought for the present study. Three-body wear will not be considered here. More detailed information on three-body tribological systems can be found, in particular, in the proceedings of the 22nd Leeds-Lyon conference in 1995 [4], which dealt with this topic.

3.1. Abrasion, erosion and cavitation wear

Wear by abrasion, erosion and cavitation is produced by contact between a particle and the surface of a solid. These types of wear are due to the action of an antagonistic material in different states: solid, particulate, liquid or gaseous.

__Abrasion wear__

Abrasive wear is caused by contact between a roughness and the surface of a solid. The roughness responsible for the wear may be a peak on the opposing surface, but it may also be debris from a surface or, more generally, a third body introduced into the contact. Damage to the surface takes the form of striations parallel to the direction of sliding. Depending on the severity of the contact, the abraded surface undergoes a number of transformations. They depend on the mechanical properties and geometry of the materials in contact. If the surface of the stressed solid is considered perfectly flat, the following damage patterns can be observed [5] (Figure 1.2):

- cutting, if the abrasive is sufficiently sharp (Figure 2.a);

- repelling of surface material (Figure 2.b);

- fracturing, if the solid is brittle (Figure 2.c);

- grain removal, if the material is insufficiently homogeneous (Figure 2.d).

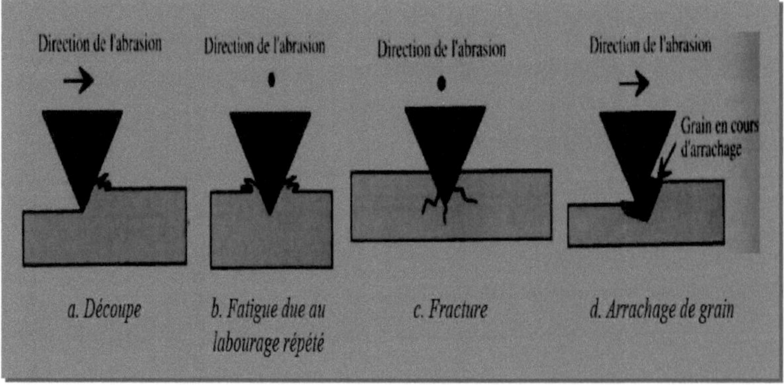

Figure 1. 2 Diagrams of abrasive wear mechanisms: cutting, fracturing, fatigue and tearing [5].

When a surface is abraded, the abrasive either forms a chip that removes material directly, in this case cutting abrasion, or forms a plastic bead on the edge of the scratch, which is removed by the repeated passage of asperities, in this case fatigue abrasion.

Fatigue on the surface of an abraded material is similar to that observed for materials subjected to volumetric stresses. The repeated passage of the abrasive causes cracks to propagate in the material from defects initially present in the material.

Fracturing of the material is due to the stresses present behind the surface or sub-surface contact. In the case of a sphere rubbing on a plane [6] gives its analytical expression. Mouginot, in his thesis and articles, gives a comprehensive account of the mechanism of contact cracking [7].

Grain removal in a material is relatively rare, and is mainly found in ceramics. This mechanism causes extremely large material removals when the cohesion between grains is low and grain size is large.

The study of material flow as a function of stress conditions provides a better understanding of abrasive wear. It depends on the mechanical properties of the material and the geometry of the indentor used (the latter imposing a deformation and a deformation speed). Measuring the friction factor also makes it possible to estimate energy dissipation in the system during the scuffing test, which is why, to model scuffing, quantities such as abrasion energy or scratch hardness are introduced.

In practice, at least two modes of action always coexist on the same surface, with the proportion of cutting generally remaining fairly low (10 to 20%) [8].

__Erosion wear__

Erosion wear is caused by the impact of solid or liquid particles against the surface of a solid. This wear mechanism depends on the properties of the materials involved, the angle of impact, the impact velocity and the particle size.

As previously, several types of erosion are observed [3] (Figure 1.3):
- cutting (Figure 3.a) ;
- fatigue (Figure 3.b)
- repoussage or cracking (Figure 3.c);
- fusion (Figure 3.d) ;
- fusion and other phenomena (Figure 3.e);
- erosion on an atomic scale (Figure 3.f).

Low-angle erosion results in the removal of material through the formation of chips.

As the angle increases, at low speeds, repeated impacts can cause cracks to propagate slowly through the material. At higher speeds, plastic deformation occurs, and the resulting bead is then removed by the impact of other particles. Under the same conditions, depending on the properties of the eroded material, cracking can lead to material removal.

At high speeds and large angles of incidence, the particle can fuse with the eroded material. In extreme cases - such as the collision of a meteorite with a planet - and for particular materials, the incident particle even fuses with the eroded solid, causing debris to be thrown up in the vicinity of the impact.

At the atomic scale, the impact between an atom and an atom of a crystal can cause the latter to be removed.

Erosional wear therefore has many similarities with abrasive wear.

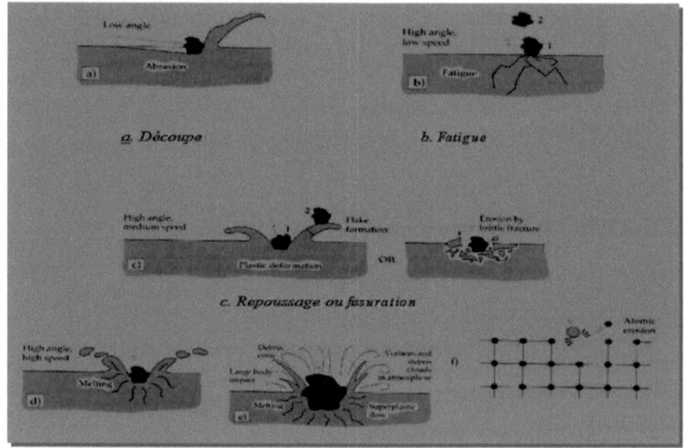

d. Fusion e. Fusion and other phenomena f. Atomic erosion

Figure 1. 3 Schematics of erosive wear mechanisms: cutting, fatigue, embossing or cracking, melting, fusion and other phenomena, and atomic erosion [3].

<u>Cavitation wear</u>

Cavitation occurs when a liquid reaches saturation vapor pressure near a surface. This creates a cavity that cannot remain stable in the medium. The cavity implodes, producing a shock wave and very high stresses on the surface, often resulting in crater-like facies.

3.2. Fatigue and delamination wear

The repeated passage of a solid over smooth or rough surfaces causes fatigue cracks to propagate and material to pull away. Cracks initiate at defects in the material. For metals, [9] explains that the initiating defects are not located on the surface or extreme surface of the material, as they are eliminated by the action of stresses on the surface. In addition, an oxide layer with higher mechanical properties protects the surface. This means that cracks initially propagate in the sub-layer, only to rise to the surface and cause the removal of wear debris and delamination wear (Figure 1.4).

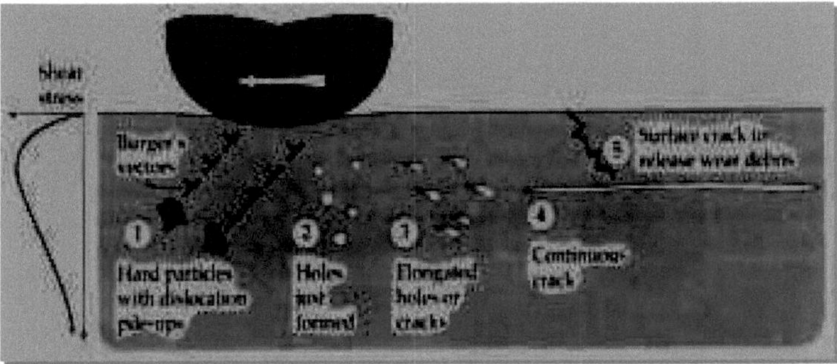

Figure 1. 4 Schematic diagram of defects causing delamination wear (after [9]).

3.3. Adhesive wear.

Adhesive wear is characterized by a high rate of wear and an unstable coefficient of friction. The wear particles are small and often transferred to the other solid in contact. This wear is due to the attractive forces existing between the atoms of the solid surfaces. When the surfaces in contact are separated by tangential or normal motion, if the forces of adhesion are greater than the strength of the material, material pull-out occurs (Figure 1.5).

Figure 1. 5 Diagram of adhesive wear. The material adheres to the antagonistic surface, stripping it of material [3].

It is possible to estimate the order of magnitude of the average diameter [10] of the wear debris thus formed. If two materials, 1 and 2, adhere to each other, (Figure 1. 6) the energy at the interface (Surface 12) is :

$$W =_{12} \frac{\pi}{4} d^2 w$$

where :

- d is the contact radius between the two solids ;

- w is the adhesion energy per unit area between the two solids in contact.

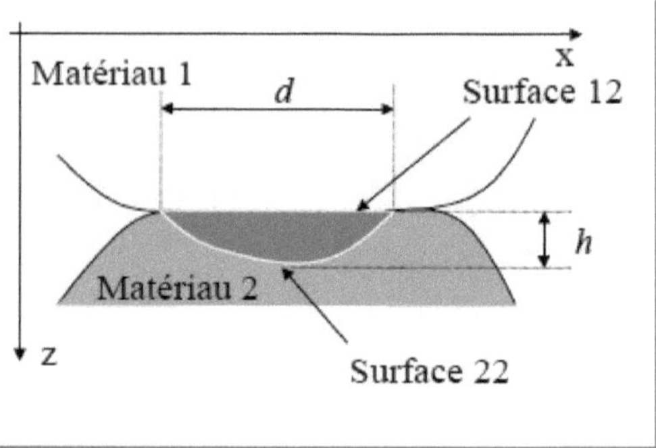

Figure 1. 6 Diagram of adhesive contact between two solids 1 (white) and 2 (grey).

The darker grey area represents the material 2 removed by adhesion. Surface 12 represents the boundary between solid 1 and solid 2. Surface 22 represents the boundary between "healthy" material 2 and debris from material 2. D is the average diameter of the debris and h is its maximum depth.

3.4. Other types of wear.

Weld wear

When two solids are in motion relative to each other, heating occurs in the contact, which can cause local welding between the materials involved, resulting in wear. This mechanism is the most severe expression of adhesive wear.

Corrosive and oxidative wear

Corrosive and oxidative wear is due to the effect of chemical reactions between the worn surface and its surrounding environment.

Fretting wear

Fretting is a low-amplitude oscillatory movement that can occur between two surfaces in contact [11]. This generates various forms of degradation that can lead to loss of functionality. The contact is often partial sliding, i.e. the two solids have no relative movement in the central part of the contact, but slide on a ring at the periphery of the contact [12]. Stress cycles cause materials to crack and debris to form, which can accelerate wear when confined within the contact.

4.   The benefits of thermal spray surface coating.

Thermal spray coatings are widely used to give or improve a specific surface quality or function of a part. Different spraying methods result in different coating characteristics, and they are mainly applied for improved resistance to corrosion, heat and wear.

5.   Thermal spraying process.

5.1. General principle.

Thermal spraying encompasses all processes whereby a filler material is melted or brought to a plastic state by a heat source, then sprayed onto the surface to be coated, where it solidifies.

As a result, the base surface does not undergo any melting.

Deposit adhesion is mechanical.

Figure 1.7 shows the general principle of thermal spraying: the material to be deposited, in powder, wire, bead or rod form, is totally or partially melted in a heat source (flame, electric arc, plasma). A carrier gas is used to atomize the material, and transport the droplets thus formed to the surface to be coated [13].

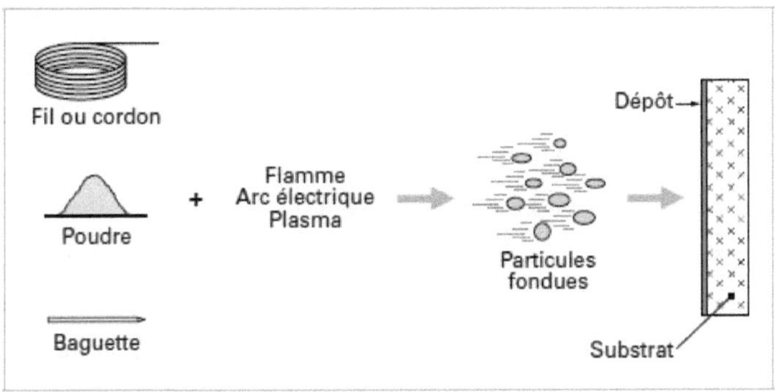

Figure 1. 7 Basic principle of thermal spraying [13].

5.2. The different projection methods.

There are many different thermal spraying processes, differentiated mainly by the energy source used (combustion, electrical discharge). The three main ones are **flame, electric arc and plasma spraying**. As the heat transfer between the sprayed particles and the substrate is relatively low, the coated part does not normally undergo any deformation, structural change or major alteration [14].

5.2.1.   Flame projection :

Flame spraying is a technique widely used throughout the world, due to its simplicity and low equipment cost. Flame guns are mainly used to spray materials in powder, wire or rod form. The principle is based on the introduction of the material to be deposited into the flame [figure 1.8]. The main enthalpic gas mixtures are acetylene/oxygen, hydrogen/oxygen or propane/oxygen. They enable stoichiometric flame temperatures of 3200°C, 2700°C and 2400°C respectively to be achieved. The projectile undergoes thermal and kinetic exchange in the flame, and is therefore melted and accelerated before being deposited in fine droplets ("splats") on the surface to be coated [15].

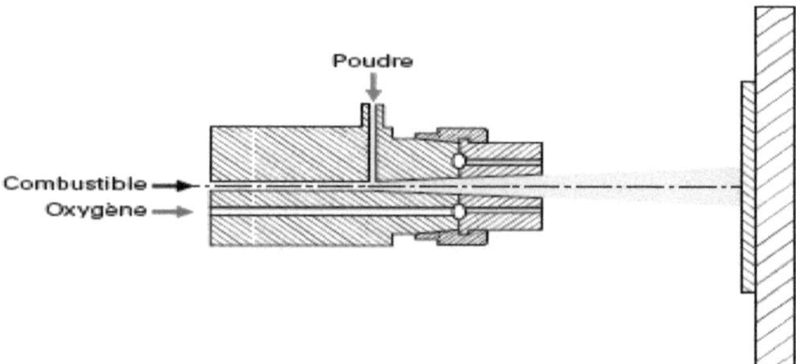

Figure 1. 8 Schematic diagram of flame projection.

The technique offers a wide range of solutions for problems such as wear, corrosion, thermal protection, etc., via both spraying and rejection treatment of the resulting coatings to improve their properties. However, it remains limited to materials with melting temperatures below that of the flame. Particle velocities vary between 30 and 150 m/s, resulting in relatively high porosity rates (between 5 and 10%) and limited adhesion. Deposition rates vary from a few kilograms per hour for powder spraying to several tens of kilograms per hour for wire spraying [16].

The advantage of this technique over other spray processes is its low cost and ease of use. The presence of large porosities in the deposit, due to the low kinetic energy of the particles and the low ejection velocity of the gases, limits the use of this technique in certain fields [17].

In general, 3 techniques can be distinguished: flame spraying, hypersonic flame spraying and detonation gun spraying. In addition, for flame spraying, a distinction is made between "powder flame" and "wire flame" spraying, on the basis of material conditioning.

5.2.1.1.   "Powder Flame

The filler material is introduced in powder form into the oxy-fuel flame, where it is accelerated by the kinetic energy transmitted mainly by the carrier gas or combustion blast gas. Torches are equipped with a powder storage container, either integrated into the torch or separate from it. The powder to be sprayed may be a pure metal, an alloy, a composite, a cermet or certain low-melting ceramics. The melting temperature of the materials used must be less than 0.6-0.7 the flame temperature (Tmelt = 0.6-0.7 Tflame).

Due to the low impact velocity of the projected particles, of the order of 30 to 50 m/s, the deposits obtained have low adhesion (of the order of 20 to 40 MPa) and relatively high porosity (10 to 20%). The materials most commonly used are self-melting alloys (Ni or Co-based alloys containing boron and/or silicon) and low-melting alloys. In the case of self-melting alloys, post-projection heating is used to obtain the deposit's refusal and bonding to the substrate. The remelted deposits are dense and virtually pore-free. Moreover, oxides are eliminated thanks to boron and silicon, to which oxygen preferentially binds and which then diffuse towards the surface as a result of the temperature gradient [18, 19,20].

5.2.1.2.   " Flamme fil

The material is packaged in wire or rod form and drawn into the flame by rollers positioned at the rear of the torch. These rollers are driven by an electric or pneumatic motor. Once the end of the rod or wire has been melted in the flame, a stream of compressed air atomizes it and sprays the particles onto the substrate surface at speeds of up to 150 m/s. Hourly deposition rates vary from 1 to 30 kg/h, depending on the type of material sprayed and the diameter of the wires or rods used. Material melting temperatures can reach 0.95 of flame temperature (Tmelt = 0.95 Tflame). Deposit thickness varies from a few tenths of a millimeter to a few millimeters. This technique can be used to spray metals (Zn, Al, Cu, Sn, Ni, Mo), steels and Zn-, Cu- and Ni-based alloys in wire form, and some ceramics (Al2O3, Al2O3 TiO2, ZrO2 with stabilizers) in bead or rod form.

5.2.1.3.　Hypersonic flame: High Velocity Oxyfuel Flame (HVOF) .

This process involves the pressurized combustion of a gaseous (propane, propylene, acetylene, hydrogen) or liquid (kerosene) fuel with oxygen or, in the case of kerosene, air. Combustion gases enter the combustion chamber at pressures of up to 1 MPa, slightly raising the combustion temperature [19]. The chamber is cooled by circulating water. The flame is then accelerated through a nozzle, reaching supersonic velocity at its exit. The powder to be sprayed is either propelled under pressure in the axis of the jet (injection upstream of the nozzle) or injected in the vicinity downstream of the nozzle at a pressure close to atmospheric pressure. Particle velocities can reach 600 to 700 m/s, and the deposits produced have good adhesion (70 to 100 MPa) and low porosity (2%). The materials sprayed are mainly cermets (WC-Co, Cr3C7-NiCr), metals and self-fusing or non-self-fusing metal alloys [21, 22].

5.2.1.4.　Detonation cannon.

In this process, the material is projected by the detonation energy of a gaseous mixture, most commonly an oxygen-acetylene mixture. The mixture is injected into the gun at the same time as the filler powder, conveyed by an inert gas. The gun consists of a long cylinder closed at one end and cooled with water. The detonation wave, created by a spark in the detonating mixture, heats and accelerates the particles, injected halfway down the barrel, to the barrel outlet, where they are ejected at high velocity (up to 900-1300 m/s) towards the surface of the substrate to be treated. Between each shot, the tube is swept by a jet of neutral gas (nitrogen). A maximum of ten shots can be fired per second. The high kinetic energy of the particles on impact with the substrate builds up a dense, adherent coating (porosity less than 1% and adhesion greater than 80 MPa). The materials generally used are chromium or tungsten carbides with a metal binder (Ni, Co) [19, 20].

5.3. Arc spraying between two wires :

The principle of arc-spraying involves sparking an electric arc between two consumable wires, thus melting the material. A jet of compressed air sprays the molten metal onto the substrate [Figure 1.9]. The temperature and high kinetic energy acquired by the particles as a result of the compressed air jet enable deposits with low porosities and good adhesion to be obtained in thicknesses of between 0.2 and 3 mm. The limitations of this process lie in the nature of the materials to be sprayed, which must be conductive (essentially metals), in the coarse structure of the deposits and in their relatively high oxidation rate due to the transport of the particles by compressed air.

This process is very simple to implement, and particularly well suited to the reliability and reproducibility requirements of medium and large-scale production [14].

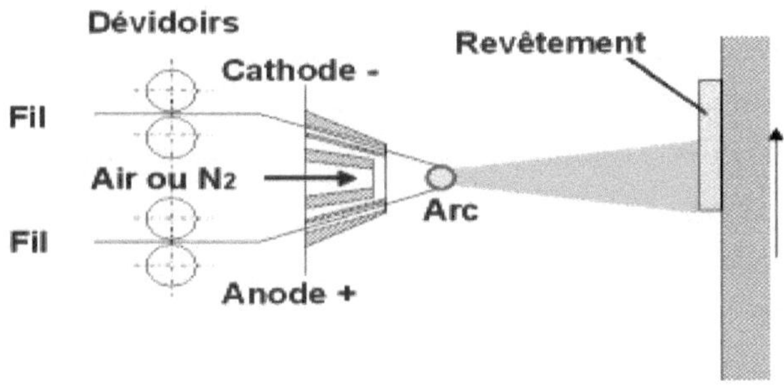

Figure 1. 9Arc spraying [14].

Source : http://www.wear-management.ch

5.4. Plasma projection :

5.4.1. Plasma.

Plasma is considered to be the 4eme state of matter (solid, liquid, gas and plasma). It is an ionized gas made up of molecules, atoms, ions and electrons, all of which are electrically neutral. An Ar/H plasma$_2$ will thus be made up of the following species: Ar, H, H+, Ar^+ , e-.

To generate a plasma, the three fundamental elements are :

➢ a power source (DC generator).

➢ a gas ionizing discharge (high-frequency or high-voltage generator).

➢ a coupling ensuring contact between two electrodes via the plasma gas.

For thermal spraying, two properties determine the speed and temperature of the particles sprayed:

✓ thermal conductivity, which determines plasma-particle transfer and therefore the state of particle fusion on impact with the substrate.

✓ viscosity, to reduce air ingress into the plasma jet, increase jet length and avoid chemical reactions of the particles (oxidation in particular) [13].

5.4.2.   Principle of plasma projection.

The principle is based on the introduction of solid particles into a plasma so that they can be melted, accelerated and finally deposited on the substrate.

The electric discharge is generated between the anode and cathode in the spray torch by a direct-current generator. The plasma gas, generally a mixture of argon and hydrogen or helium, is injected between the electrodes where the electric arc is produced. The gas undergoes molecular dissociation (in the case of hydrogen) and/or partial ionization, resulting in strong expansion and expansion of the plasma jet at the nozzle outlet [Figure 1.10].

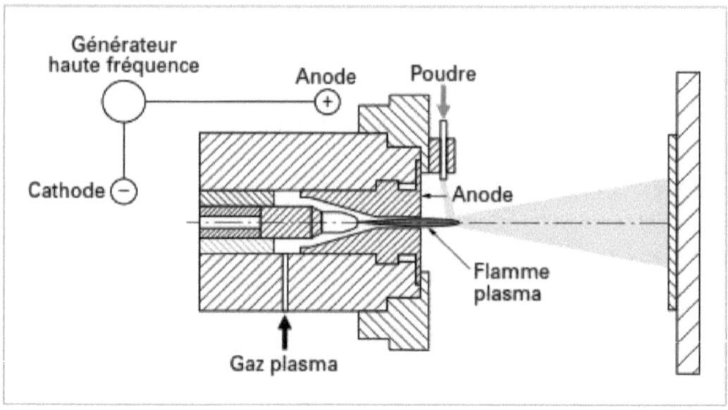

Figure 1. 10 Schematic diagram of plasma spraying [13].

5.4.3.   The different plasma spraying processes.

A wide range of plasma spraying techniques are available. The processes listed below are by no means exhaustive [23]:

✓ A.P.S.: Atmospheric plasma spraying ( );

✓ V.P.S.: Vacuum plasma spraying ;

✓ R.P.S.: Reactive plasma spraying (under reactive atmosphere; for example, in the presence of nitrogen to nitride a material);

✓ I.P.S.: Inert plasma spraying (when the atmosphere is controlled; for example, air is replaced by a neutral gas);

✓ H.P.P.S.: High pressure plasma spraying (in the presence of a slight increase in pressure to promote heat transfer).

5.4.4. Some mechanical properties of plasma sprayed coatings.

The mechanical properties of plasma thermal spray coatings depend on the process used. Tensile strength can range from 30 to 70 MPa and porosity from 1 to 10%.

5.5. Applications.

Manufacturers' objectives are essentially to reduce costs and improve performance. Applications are diverse and varied.

Table 1 gives some examples of applications and materials used in the thermal spray sector [23].

	Exemple de pièce	Exemple de matériau projeté
Barrière thermique	Réacteur	$ZrO_2 - X$ (APS)
Réparation	Axe, arbre, carter	Sous-couche + matériau de base (FS)
Résistance à la corrosion	Cuve	316 L (HVOF), NiCrAlY, NiCr
Résistance à l'usure	Arbre, aubes de turbine	WC-Co-Cr (HVOF), Al_2O_3 (PS)
Gain de poids	Chemise de moteur	Acier sur un alliage d'aluminium (APS, TWEA)
Biocompatibilité	Prothèse	Hydroxyapatite sur $TiAl_6V_4$ (APS)

Table 1Examples of thermal spray applications [23].

6.   Conclusion.

In this chapter, we have outlined the importance of thermal spraying in many industrial sectors, particularly the mechanical engineering industry. Thermal spraying is becoming more and more widespread, as coated parts have specific characteristics and a manufacturing cost advantage. The principle of thermal spraying and the different thermal spraying techniques have been presented in this section... However, it is essential to know the adhesion of the coating to the substrate, and to be able to quantify it quickly and reliably.

CHAPTER II: Adhesion of coatings

1. Coating.

1.1. Definition.

Dry tribological coating is a technique developed for fasteners and components subject to mechanical stress (screws, nuts, washers). It is a thin, non-electrolytically deposited coating with lubricating properties, providing additional protection against corrosion. The coating consists of a composition containing fluoropolymers and submicroscopic organic particles of a solid lubricant, dispersed in carefully selected mixtures of synthetic resins and solvents. The coating, known as AFC (Anti-Friction-Coating), forms a thin, smooth layer that corrects surface irregularities, reducing friction even under high stress and extreme working conditions. The synthetic resin, meanwhile, provides enhanced protection against corrosion [24].

1.2. Coating process.

The coating is applied using spray guns, either individually (in an automatic passing machine or manually), or in bulk in rotating drums. The sprayed layer is then cured in the oven, where it acquires its outstanding adhesion and corrosion protection properties. Depending on the specifications required, the layer thickness varies between approx. 5 and 12 µm.

Figure 2. 11 Flame projection. http://tss.asminternational.org/portal/site/TSS

1.3. Mode of action.

A thin, dry film of lubricant firmly adhering to the base forms after the antifriction coating has cured. This film acts as a lubricating and separating layer, reducing friction and wear between rubbing counterparts. The application of a special primer for the first coat considerably reduces or even eliminates wear on the dry tribological coating. The service life of the applied layer depends on a multitude of product characteristics, including the wear resistance of the binder system, its elasticity and adhesion to the surface of the components.

1.4. Main features.

- Excellent, low-dispersion friction values as the basis for every screw connection.

- Environmentally-friendly, dry coating with remarkable ease of use.

- High assembly safety during manufacturing and maintenance.

- Cost-effective assembly/disassembly, with an overall reduction of up to 30% in process costs.

1.5. Advantages of the coating process

- High protection against fretting corrosion, especially when in contact with aluminum and stainless materials.

- Good separation of building materials thanks to the organic layer and, consequently, low electrical conductivity.

- Highly resistant to fuels, hydraulic fluids and cleaning solvents.

- Excellent sliding properties and, consequently, optimum protection against seizing/clogging of assembly elements; easy detachment of the assembly after use.

- Lubrication assured.

- No adverse influence on the mechanical properties of high-strength joining elements; sufficient ductility for elastic deformation.

- With appropriate surface treatment, no hydrogen embrittlement occurs during dry coating.

- Additional colored lacquering available if required.

- Sufficient adhesion for microencapsulation after appropriate pre-treatment (limit values prescribed by applicable standards can only be met under certain conditions).

2.   Coating adhesion.

2.1 Definition.

It is   important to differentiate between the two French words "adhérence" and "adhésion", which are both translated by the word "adhésion" in English, but which do not have the same meaning.

Adhesion represents the surface bonding of one material to another. This bond is achieved by forces located at the interface between the two materials [25]. These forces can be of different natures. A distinction is made between Van-Der-Waals forces2 (physical interaction) and chemical or diffusion bonding forces (metallurgical interaction) [26, 27,28]. Adhesion characterizes the resistance to rupture of the interface between two materials [29].

Adhesion and adhesion are therefore two different but complementary concepts. Adhesion characterizes the strength of an assembly of two materials held together by adhesion.

2.2 The challenges of measuring grip.

We saw in the previous chapter that thermal spraying is a technique for creating a coating with the aim of modifying the surface mechanical properties of a part. Adhesion, although not directly part of the desired properties, is an essential parameter in the reliability of the deposited coating. Indeed, what good would the best-performing coating be if it were to detach from the substrate during the first loading cycle? The designer must therefore take care to create strongly adherent coatings, especially in the case of parts subjected to high loads. The next step is to determine the adhesion of the coatings, using simple, fast and reliable techniques.

2.3 How to determine adhesion.

Adhesion can be assessed in a number of ways, for example in terms of force or workload. According to Rickerby [28], the measurement should ideally meet a number of criteria. It should :

> ➢ be non-destructive.
> ➢ be easily adaptable to systematic testing of parts with complex geometries.
> ➢ be relatively simple to perform and interpret.
> ➢ enable standardization and automation.
> ➢ be reproducible.
> ➢ be directly related to the reliability of the coating in its specific application.

2.4 Mechanisms of adhesion of deposits to the substrate.

Adhesion ensures the formation and cohesion of the interface between two solids (the deposit and the substrate). It is induced by the forces of attraction between the materials. Adhesion mechanisms can be of several types:

✓ mechanical: adhesion is ensured by mechanical anchoring of the material, which penetrates the substrate's microcavities and surface irregularities prior to solidification. It has been shown experimentally by a simple example that a polished material has a lower degree of adhesion than a sandblasted material [30, 31]. Good mechanical anchorage requires a judicious choice of surface roughness and depends on both the adhesive and substrate wettability [32], as well as the contact surface between substrate and adhesive. Good deposit/substrate surface affinity therefore limits the formation of defects linked to trapped air and crack propagation [33, 34, 35, 36]; this mechanism is the most frequent, particularly in the case of ceramic deposits .

✓ chemical: this involves "chemical" adhesion to the point of forming true bonds resulting from intermolecular or interatomic interactions (e.g. covalent or ionic) at the interface. All these different forms of bonding contribute to the good adhesion of the substrate/deposit system. This mechanism is also observed in the case of high-temperature heat treatment during coating production, or in the case of an exothermic reaction that can promote the formation of metallurgical bonds at the interface.

Nevertheless, bonding with the substrate can be enhanced if the latter has been prepared in advance. Unsatisfactory preparation can have a detrimental effect on the adhesion of the coating to the substrate. In fact, the nature, temperature and surface preparation of the substrate influence heat transfer [37], through changes in the substrate's mechanical properties... etc.

The substrate must therefore be pre-treated to adapt its roughness, so as to create mechanical anchoring points for the sprayed particles. However, it is important not to create too great a roughness, which would reduce the cooling and spreading rates of the particles by locally increasing the thermal contact resistance [38].

3.   Thermal spraying operating parameters.

Thermal spraying is characterized by a strong interaction between operating parameters and the properties and characteristics of the deposits produced. Operating parameters can be divided into five categories (figure 2.12):

- ❖   Parameters relating to filler materials.
- ❖   Parameters for injecting the material into the gas jet.

❖   Jet energy parameters.

❖   Kinematic parameters.

❖   Deposition/substrate interaction parameters.

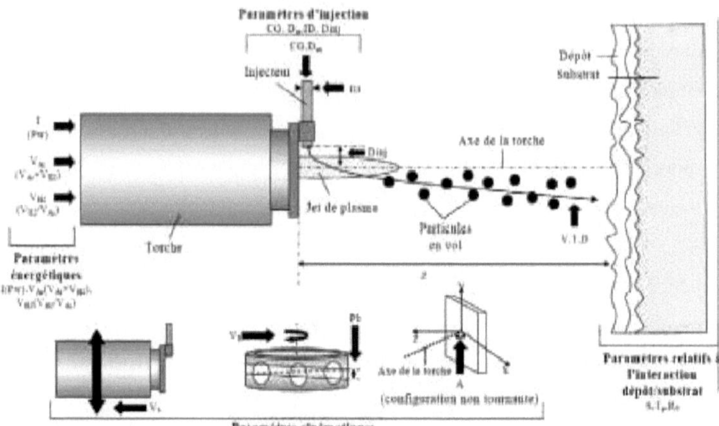

Paramètres d'injection	
Dm : débit massique de la poudre [g.min⁻¹]	CG : débit du gaz porteur [Nl.min⁻¹]
ID : diamètre de l'injecteur [mm]	Dinj : distance injecteur/axe géométrique de la torche [mm]
AI : angle d'injection	
Paramètres énergétiques	
I : intensité du courant d'arc [A]	Pw : puissance électrique [w]
V_{H2} : débit d'hydrogène [Nl.min⁻¹]	V_{Ar} : débit d'argon [Nl.min⁻¹]
$V_{H2}+V_{Ar}$: débit total [Nl.min⁻¹]	V_{H2}/V_{Ar} : taux d'hydrogène [%]
Paramètres cinématiques	
Z : distance de projection [mm]	A : angle de projection [°]
V_b : vitesse relative torche/substrat [m.s⁻¹]	V_r : vitesse de balayage [mm.s⁻¹]
P_h : pas de balayage [mm]	
Paramètres relatifs à l'interaction dépôt/substrat	
S : nature de substrat [-]	Tp : température de substrat [°C]
R_s : rugosité de la surface de substrat [um]	

Figure 2.12 Categories of operating parameters [39].

Operating parameters are strongly coupled to each other, as several studies have shown [39, 40,41]. In this paper, we will concentrate on the kinematic parameters.

The kinematic parameters concern (figure 2.13) :

❖   the torch's trajectory,

❖ the relative speed of torch movement with respect to the substrate,

❖ projection distance,

❖ projection angle,

❖ Scanning pitch (distance between two adjacent passes of the torch).

These parameters can influence deposition yield, substrate temperature, lamella and deposit morphology, and deposit quality [42,43].

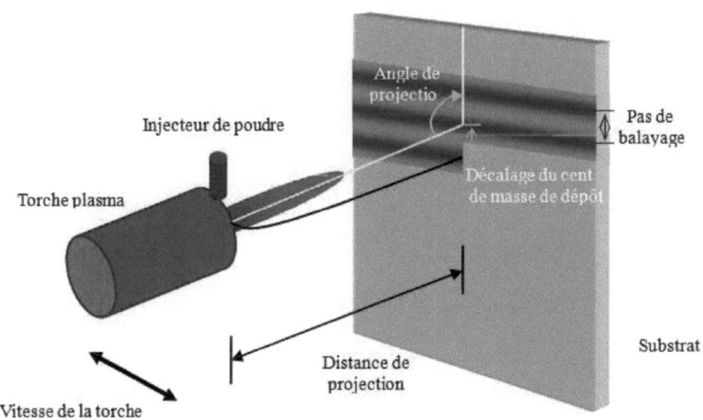

Figure 2. 13 Kinematic parameters.

3.1 Relative speed torch - substrate .

The relative torch-substrate speed represents the difference in displacement speed between the torch and the substrate (Figure 2.13). Along with powder flow rate, sweep pitch and number of passes, this speed is one of the most influential parameters on the mass distribution of material sprayed onto the substrate [41].

The quality of a deposit generally lies not only in its structural homogeneity (low porosity, few voids, little oxidation), but also in its homogeneity of distribution [39]. Obtaining a constant thickness is a guarantee of material savings (compliance with a minimum rib) and reduced post-processing (machining to dimension).

The thickness of material deposited on a surface is a function of the torch illumination speed relative to the substrate. As a first approximation, obtaining a homogeneous deposit in terms of thickness requires a constant impact speed over the entire substrate, if the other parameters remain constant. Hence the need to define a torch displacement speed that ensures a constant illumination rate.

Thermal spraying, like painting and gluing, is one of the so-called "continuous" processes. In other words, the speed of illumination must be constant, and the movement must be adapted to the substrate profile.

While the movement of an industrial robot is geometrically precise, the kinematic and dynamic characteristics of the tool movement are variable during the execution of the trajectory. There is no guarantee that the specified speeds will be respected, especially at high accelerations. It is therefore essential to generate trajectories that guarantee a constant illumination speed.

On the other hand,   the scanning speed needs to be altered during projection on many rotating parts of variable cross-section (a configuration often encountered in industrial applications) (Figure 2.14).

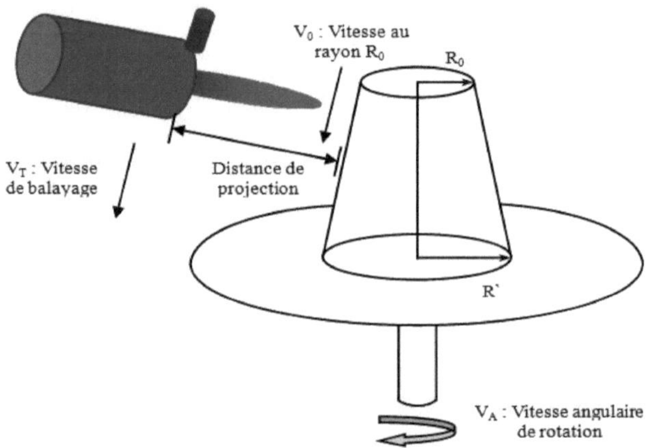

Figure 2.14 Projection on a rotating sample.

The speed of the torch movement, coupled with the rotation of the sample, defines the scan pitch, which measures the degree of overlap of the deposited material indentations. In addition, the linear speed of the torch is a function of the angular speed of the part and the local radius (Eq.1).

$$V_L = 2\,\pi\,R.V_A \text{ (Eq.1)}$$

where VL is the linear speed [mm.s-1] and VA is the angular speed [rev.s-1]. In this situation, the torch speed must be adapted to the radius R, which means that the sweep

28

pitch must decrease as the projection speed increases. As a function of the linear velocity at *R0*, torch speed therefore varies with radius *R* (Eq.2).

$$V_T = \frac{R0}{R} \cdot V_0 \text{ (Eq.2)}$$

The torch speed must therefore be carefully controlled during the thermal spraying process.

3.2 Projection distance

The throw distance is the distance between the last geometric plane of the torch nozzle and the substrate (Figure 2.12, Figure 2.13, Figure 2.14). Ideally, all particles impacting the substrate are completely melted. However, because of their different sizes, injection, shapes, behaviors, and the significant velocity and temperature gradients in the plasma jet, the particles arrive at the substrate in different states of fusion [44]. This is also the most easily adjustable parameter for varying the state of the particles on impact with the substrate.

An optimum spray distance value should establish a relatively equal temperature value between small and large particles prior to their impact on the substrate [45,46]. Spraying distance controls deposition properties such as residual stress rate, porosity, infondus rate and spraying yield [47,48,49].

A very short spraying distance leads to excessive heating of the substrate due to the thermal fluxes transmitted. On the other hand, too great a distance can lengthen particle trajectories and thus cool them by convective/radiative transfer, rapidly influencing deposition yield, porosity rate and adhesion to the substrate [50,51] (Figure 2.15).

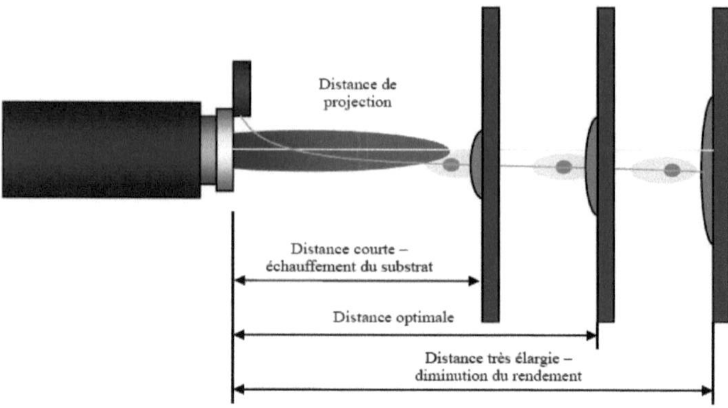

Figure 2.15Influences of projection distance

This is why, in order to guarantee homogeneity in the intrinsic characteristics of the deposit, a constant projection distance must be maintained along the entire trajectory.

3.3 Projection angle

The projection angle is commonly defined as the interior angle between the geometric axis of the torch and the tangent to the surface to be coated (Figure 2.12, Figure 2.13). The impact of particles on the substrate can be classified into two groups: normal impact and inclined impact.

The spray angle has a significant influence on the profile of the deposit formed. Studies [52, 53,54] have shown that the height of the deposited imprint decreases with decreasing spray angle, thus also reducing the spray yield as a result of increasing particle bounce on the substrate (Figure 2.16). This decrease is generally accompanied by an offset of the indentation [55].

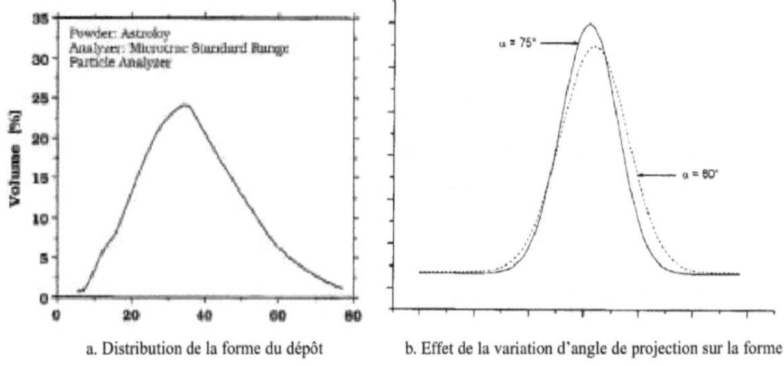

a. Distribution de la forme du dépôt   b. Effet de la variation d'angle de projection sur la forme

Figure 2. 16 Deposit shape distribution and projection angle effect [41].

In addition to these macroscopic effects, a remarkable effect on porosity rate and morphology has been demonstrated, corresponding to an increase in porosity rate as the projection angle decreases [56,57]. This generally causes a loss of adhesion and a decrease in hardness. The "ideal" projection angle for maximum projection efficiency would therefore be the normal angle. Several studies [53,55] have also highlighted the fact that surface roughness increases with decreasing spray angles.

The need for perfect control of torch orientation is therefore obvious, and an important parameter to take into account when thermal spraying.

3.4 No sweeping

The scanning pitch is the distance between two consecutive passes of the spray torch. In the case of thermal spraying (e.g. APS), the characteristic dimension of the impact zone of the sprayed particles at the spraying distance varies between 8 mm and 20 mm, depending on the material and operating conditions. The value of the scanning pitch naturally influences the thickness and shape of the final deposit.

An optimum scan pitch value should establish a fine texture of the deposit surface and prevent the substrate temperature from rising too quickly (Figure 2.17), as a fine pitch provides a more homogeneous deposit but causes more local heating of the substrate. In general, a scan pitch value ranging from 5 mm to 15 mm is often used in the APS process.

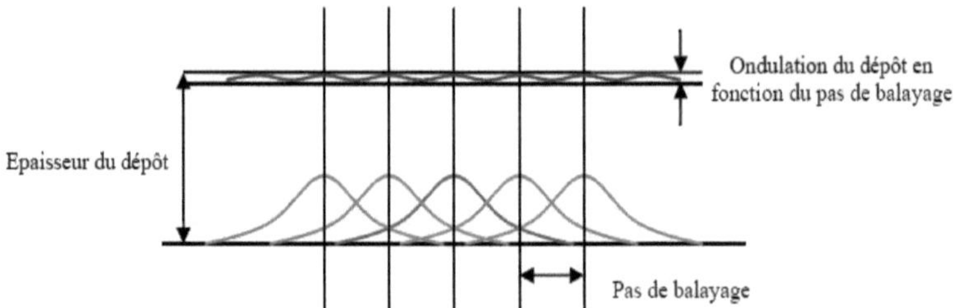

Figure 2. 17 Deposit surface texture

The right choice of sweep pitch is therefore important to obtain a suitably shaped coating. This operating parameter must therefore be well established before thermal spraying and well controlled during spraying.

CHAPTER III: EXPERIMENTAL PROCESS

1. Thermal gas spraying :

1.1. Introduction

Gas thermal spraying (oxyacetylene) is one of the oldest and longest-established spraying processes. Much less widely used today, it still has the triple advantage of versatility, ease of use and low equipment costs. It is ideally suited to repair or construction work, or for spraying tubes and structures made of materials (such as cast iron) particularly prone to cracking, with wall thicknesses of 0.5 to 6 mm. It is also widely used for spraying non-ferrous metals and for plating or hardfacing, as well as for flame cutting, heating and straightening.

Heat is supplied by the combustion of acetylene in oxygen. The flame temperature reaches around 3,100°C. This temperature is lower than that of the arc, and the heat is less concentrated. The welder directs the flame onto the joint surfaces, which melt. Filler metal can then be added as required. The weld pool is protected from the air by the reducing zone and the secondary zone of the flame. The flame must therefore be gradually withdrawn at the end of the spraying operation.

As the flame is less concentrated, cooling is slower, which is a definite advantage when spraying steels, which tend to harden. At the same time, however, since the process is relatively slow, the heat input is greater and the risk of thermal stress and deformation increases.

1.2. Hardware

Gas-blasting equipment includes (Figure 3.18):

- ➢ gas cylinders ;
- ➢ pressure gauges/pressure reducers/regulators ;
- ➢ gas hoses ;
- ➢ flashback arrestors ;
- ➢ projection torches.
- ➢ Filler materials.

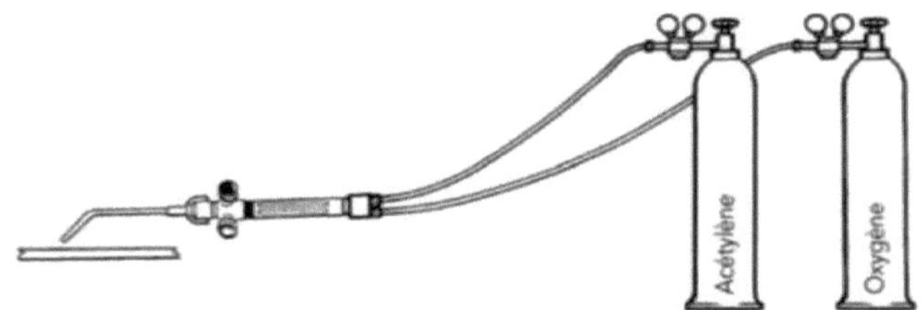

Figure 3. 18Gas-blasting equipment.

1.2.1.   Spray gases and their storage

Flammable gas cylinders must be stored outdoors or in well-ventilated areas. Specific signs must be posted outside the storage area. Acetylene and oxygen cylinders must be carefully separated.

Acetylene

Acetylene (C2H2) is the main fuel gas used for gas spraying. Table 2.1 shows its main properties in relation to other fuel gases. It consists of 92.3% carbon and 7.7% hydrogen. Its combustion in oxygen produces a higher combustion temperature than any other gaseous hydrocarbon, and its flame is far more concentrated than that of any other gas.

Gaz	Densité (kg/m³)	Valeur calorifique (MJ/kg)	Température de flamme (°C)	Vitesse de combustion (m/s)
Acétylène	1,07	48,2	3 100	13,1
Propane	2,00	46,4	2 825	3,7
Hydrogène	0,08	120	2 525	8,9

Table 2Main characteristics of combustible gases.

Acetylene is highly flammable. It forms a highly explosive mixture with air at concentrations ranging from 2.3 to 82%. Care must be taken to prevent any leakage from cylinders or pipes.

Even in the absence of air, pressurized acetylene is chemically unstable and can, under certain conditions, explode into carbon and hydrogen. It is stored in cylinders filled with a porous material saturated with acetone, which absorbs the gas at a pressure of 2

MPa. If the pressure exceeds 1.5 MPa, explosive decomposition can occur in the pipes leading from the bottle.

Oxygen

Oxygen is stored under pressure or in a liquid state. In cylinders, it is stored under 20 MPa. Large-scale users generally receive it in liquid oxygen form.

When directed at a flammable element, pure oxygen ignites easily. Connections must be clean and tight to prevent beading. They must never be oiled or greased.

1.2.2.   Pressure reducing valves

The gas is stored in a high-pressure cylinder (pressure varies according to filling). The regulating valve delivers the gas at working pressure, ensuring a constant flow rate despite variations in counter-pressure due to heating of the spray torch.

1.2.3.   Gas hoses

The gas hoses are red for acetylene and blue for oxygen. To avoid connection errors, the acetylene connection is left-hand threaded, while the oxygen connection is right-hand threaded.

1.2.4.   Flashback arrestor

Flame flashback, characterized by the snapping sound that accompanies it, is synonymous with the re-entry of the flame into the torch. It occurs when the gas flow no longer keeps pace with the flame's rate of combustion, so that the flame front moves backwards. When a flashback, recognizable by the hissing sound it produces, is prolonged, all gas valves and taps must be closed immediately.

In fact, flashback occurs when oxygen and acetylene flow back up the pipes - when oxygen enters the acetylene pipe and the two gases form an explosive mixture. A flashback arrestor fitted to the regulator stops the flame front and prevents the flame from reaching the acetylene cylinder and causing an explosion.

1.2.5.   Projection torches.

The powder flame spray gun (Figure 3.19) is a separable unit consisting essentially of :

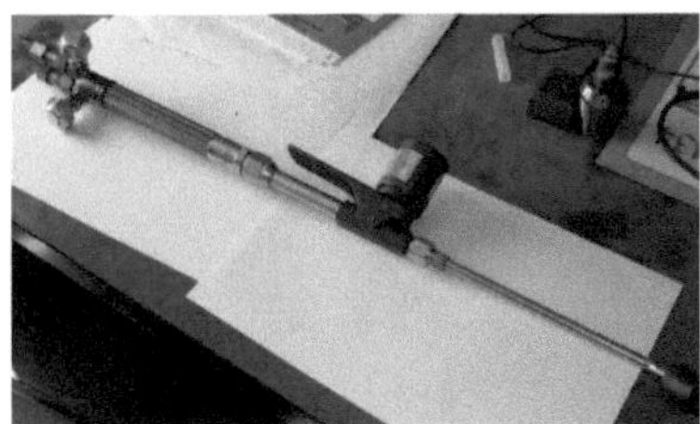

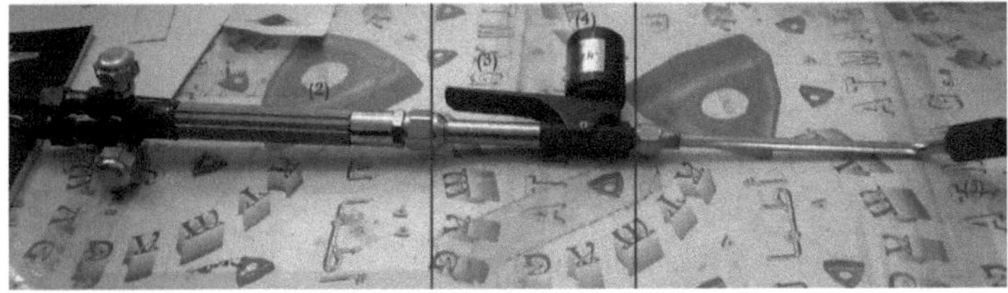

Figure 3. 19Powder flame spray gun.

❖ Two taps for loading the gun with oxygen and acetylene gases, which trigger the combustion reaction and transmit the powder to the nozzle.

❖ Gas transmission channel to the nozzle.

❖ A wrist that allows powder to escape.

❖ A body that holds the powder tanks in place.

❖ A nozzle represents a powder outlet and spray opening.

1.2.6.   Filler materials.

The filler materials available and most often used in our thermal spray tests (Figure 3.20) are :

- CW1085 carbide with hardness in the range [58-62].

Figure 3. 20Tungsten carbide powder.

2. Safety instructions.

- ➢ Before opening gas cylinders, check:
- • Waterproofing.
- • Accessories in good condition.
- • Correct installation of the system.
- ➢ Never oil or grease gas cylinder valves, holders, fittings or flashlights.
- ➢ Never allow oil, grease or other organic matter near oxyacetylene units.
- ➢ Never handle a burning flashlight without gloves and eye protection.
- ➢ Never point an oxyacetylene flame at gas cylinders.
- ➢ Never arc gas cylinders.
- ➢ Do not expose hoses to damage from flames, sparks or falling sharp objects.
- ➢ In the event of an incident, quickly close the acetylene cylinder valve, then the oxygen cylinder valve.

3. Application:

For our application, we used a rectangular steel part as shown in figure (3.21) (length 280 mm, width 40 mm and thickness 17 mm), and divided it into two parts, a smooth part and a rough part.

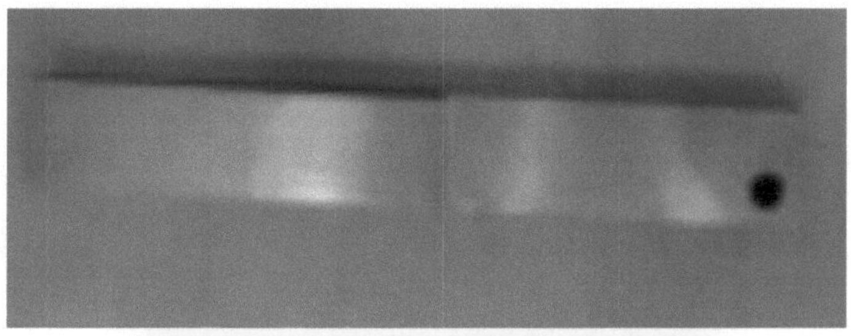

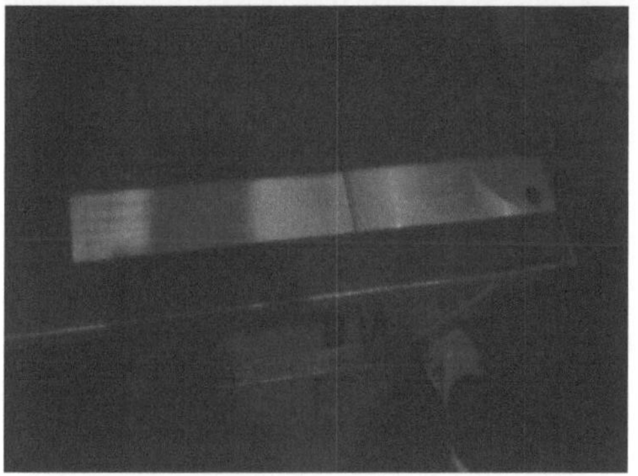

Figure 3. 21Sample processed in our trial.

In our test, we'll be working with two cylinder pressures: 4 bar for oxygen and 1 bar for acetylene, but the flowmeter is not specified. The torch-substrate distance varies from 3cm to 6cm, and the projection flame is around 3,000°C. However, the part temperature varies between 600°C and 800°C.

Figure 3. 22Flame-powder projection on a steel part.

The aim of our study is to develop an ultrasonic test, and to compare the echoes resulting from the reflections of the ultrasonic wave on the tungsten carbide-coated steel part, and to calculate the attenuation (see Appendix A.I).

3.1. Characterization methods :

The coating produced is characterized by ultrasound, one of a number of non-destructive testing methods. A 1085 ultrasound scanner and a translator generating ultrasonic contact waves were used to perform the manipulations. The translator frequency is 2 MHz.

The results found are interpreted in the remainder of this chapter.

<u>Ultrasonic Non Destructive Testing : Contact Testing</u>

Ultrasonic non-destructive testing is one of the most widely used procedures in industry today, thanks to its safety and penetrating power. The principle consists in sending an ultrasonic wave into the material to be inspected. The wave is reflected by the defect and picked up by an ultrasonic transducer. By measuring the wave's time of flight and amplitude, we can locate and, in some cases, size the defect.

Brief description of the ECHORGAPHE 1085 and its software (see Appendix A.II).

3.2. Results and discussion:

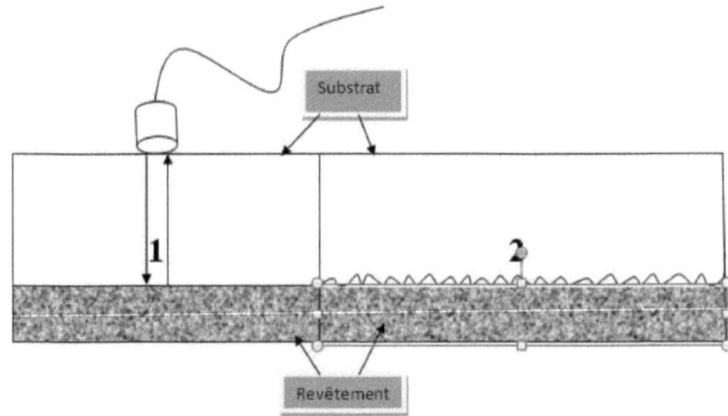

Figure 3. 23Checking a tungsten carbide-coated steel substrate with a straight sensor.

<u>The smooth part (1) before and after coating :</u>

The echoes measured before coating (figure 3.24) are somewhat distant, with a slight attenuation of the signal due to the absorption of the 2 MHz ultrasonic wave in the part and its dispersion.

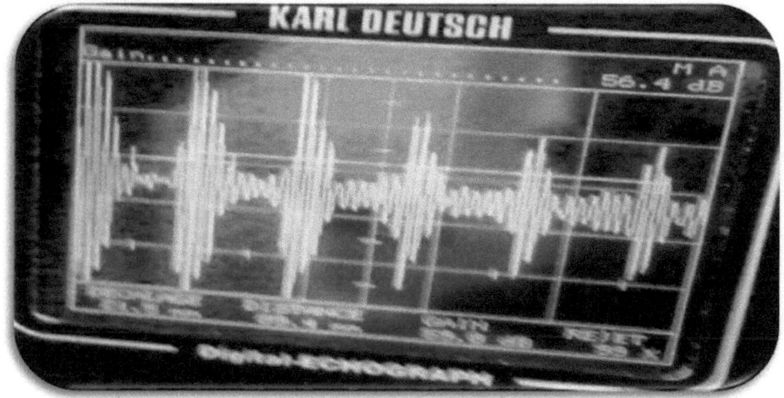

Figure 3. 24Echo before coating.

On the other hand, after the coating has been applied (figure 3.25), the same echoes have become enlarged and confused with each other after passing through a certain thickness. Attenuation of the ultrasonic signal is low.

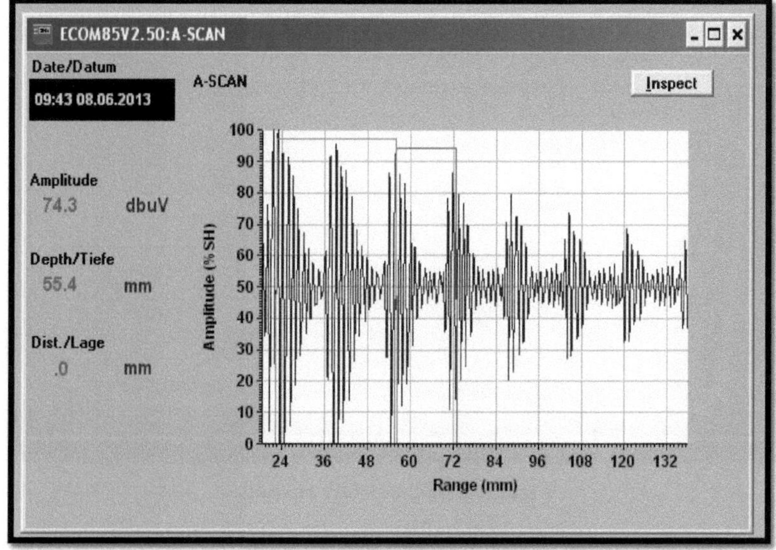

Figure 3. 25Echo after coating.

$\boxed{\Longrightarrow}$ Signals appearing after coating are due to reflection-transmission phenomena on the part-coating interface (interface defect).

Conclusion: the smoothness of the surface (1) leads to poor adhesion of the coating deposited on it, due to the lack of roughness (anchoring of the grains on the surface) at the interface.

The attenuation measured in the smooth part is generally given by :

$$\alpha = \frac{1}{2e}\ln(\frac{A_1}{A_2})$$

From figure A- scan 1 we have :

Part thickness e = 17mm

The amplitude: A_{10} = (94-50)=44db and A_{01} =342db so A_1 = 86db.

$$A_{20} = (85\text{-}50)\text{=}35\text{dB and } A_{02} = (50\text{-}10)\text{=}40 \text{ dB so } A_2 = 75\text{dB}$$

$$\text{Donc}: \alpha = 0.004mm^{-1}$$

The rough part (2) before and after coating:

The echoes are distant and clean before coating, with significant attenuation (see Figure 3.26).

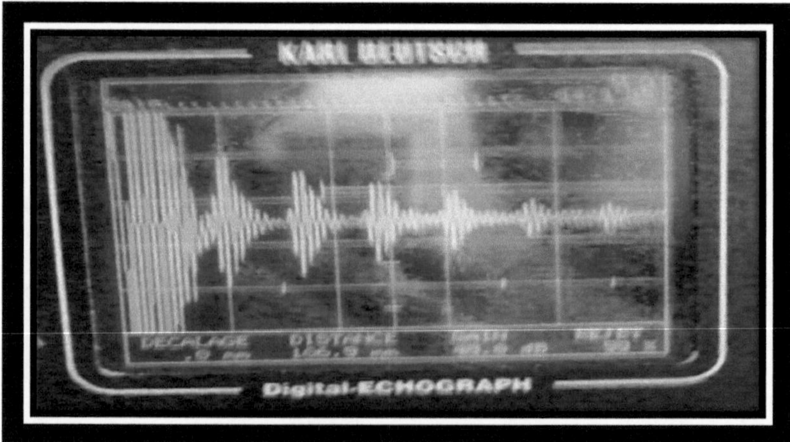

Figure 3.26Echo before coating.

After depositing the coating layer (Figure 3.27), the echoes became more disturbed and broader, but still with a high degree of attenuation (signal damping).

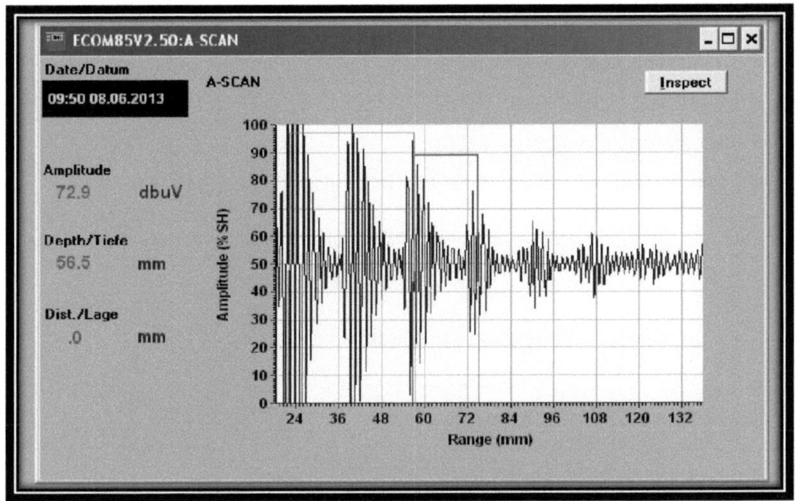

Figure 3.27Echo after coating.

<u>Remarks</u> :

- the high attenuation is due to multiple reflections on the rough surface's asperities.

- signals appearing after coating may be due to typical defects (micro-voids, impurities, unfounded particles, etc.) contained in the coating as a result of the spraying parameters, or to the interaction of the ultrasonic wave with the asperities of the rough surface.

The attenuation is: $\alpha = \dfrac{1}{2e}\ln(\dfrac{A_1}{A_2})$ From figure A-scan **2** we have :

Part thickness e = 17mm

Amplitude: $A_{10} = (96\text{-}50)\text{=}46dB$ and $A_{01} = (50\text{-}24)\text{=}26$ dB and $A_1 = 92dB$

$A_{20} = (77\text{-}50)\text{=}26dB$ and $A_{02} = (50\text{-}24)\text{=}26$ dB so $A_2 = 53dB$

So $\alpha = 0.015mm^{-1}$

3.3. Interpretation :

The difference between the attenuation coefficients of smooth and rough surfaces is generally due to absorption and diffusion phenomena within the part.

Since the problem here is the difference in surface roughness of the part, we're going to look at the phenomenon of diffusion, because a high level of roughness will increase the number of joints between grains on the surface. The smoother the surface, the less energy will be wasted, and vice versa.

The results confirm this interpretation.

The attenuation coefficient of the smooth surface

$$\alpha_{lisse}= 0.004 \text{ mm}^{-1}$$

The attenuation coefficient of the rough surface

$$\alpha_{rugueuse}= 0.015 \text{ mm}^{-1}$$

Hence :

$$\alpha_{lisse} < \alpha_{rugueuse}$$

High attenuation implies high diffusion at the surface due to the joints between the grains. So the more joints there are at surface level, the **better the material will bond to the substrate**.

3.4.   Conclusion:

The echoes measured at low frequencies do not provide a reliable characterization of the adhesion of ceramic coatings to their metallic substrate. Indeed, the coating thickness $\sim$ 2 mm and the wavelength used in this study is of the order of 3 mm, yet to be able to detect defects in the coating and control adhesion at the interface, the following condition must be met: $e > 10*.\lambda$ where e is the substrate thickness and λ wavelength.

In the case of ceramic deposits, the adhesion and cohesion of the coating are essentially of mechanical origin (anchoring) and a function of the height of the roughness peaks as shown in several previous works, so we can say in this case that the adhesion of the coating on the rough part of the substrate is much higher than that on the smooth part hence the great interest of surface preparation before coating.

Surface preparation (corundum blasting, preparation by machining or chemical pickling...) is one of the key elements in obtaining the right roughness to optimize the performance of the coating and the use of the material whose properties are to be modified. However, these treatments must be carried out a few minutes before spraying to limit the risk of oxidation.

General conclusion :

Thermal spray coating techniques form part of a wide range of surface treatment processes, the essentials of which are described in the first chapter. Plasma spray coating is one of these processes, and today it represents an industrial process offering effective solutions in a wide range of fields, including biomedical and health (implants, dental crowns, prosthetics), energy (gas turbines, alternators, combustion chambers) and transport (land vehicles, trains, aircraft and spacecraft).

We opted to focus on our main target, which is thermal spray coating, where we explained the various factors that can have a major influence on the efficiency and quality of spraying.

Through our work, we have concluded that this high-performance technique requires particular interest and advanced research to master it, while admitting the reality of its absence at the level of Moroccan companies, which leaves out an opportunity for development and irrecoverable profit.

Indeed, this technique meets the ecological expectations of manufacturing processes, as it is a dry process, which does not use pollutants during its implementation and leads to very little waste at the end.

Reference and Bibliography :

[1] G. ZAMBELLI, L. VINCENT, Matériaux et Contacts - une approche tribologique (1998), Presses polytechniques et Universitaires Romandes.

[2] Olivier BARRAU, "Etude du Frottement et de L'usure D'acier à outils de travail à chaud", Thesis, pp 12,2004.

[3] G. W. Stachowiak and A. W. Batchelor, "Engineering Tribology", Butterworth-Heinemann, Woburn, E.U., 2001.

[4] Leeds-Lyon, The third body concept: interpretation of tribological phenomena, Elsevier, Amsterdam, P.B., Tribology Series, (31), 1995.

[5] J.-M. Georges, "F friction, wear and lubrication", Editions Eyrolles / CNRS Editions, Paris, France, 2000.

[6] G. M. Hamilton, "Explicit equations for the stresses beneath a sliding spherical contact", Proc. Instn. Mech. Engrs, 197, (C), pp 53-59,1983.

[7] R. Mouginot, "Crack formation beneath sliding spherical punches", Journal of Materials Science, 22, (3), pp 989-1000, 1987.

[8] A. Cornet and J.-P. Deville, "Physique et ingénierie des surfaces", EDP Sciences, Les Ulis, France,1998.

[9] N. P. Suh, "The delamination theory of wear", Wear, 25, pp 111-124, 1973.

[10] E. Rabinowicz, "Friction, Wear and Lubrication", Massachusetts Institute of Technology, Cambridge, E.U., 1974.

[11] S. Fouvry, "Etude quantitative des dégradations en fretting", Ecole Centrale de Lyon, Mécanique, N° d'ordre 97-04 (thesis), 1997.

[12] B. D. Mindlinet H. Deresiewicz, "Elastic spheres in contact under varying oblique forces", Journal of Applied Mechanics, September, pp 327-344, 1953 .

[13] Alain PRONER, Thermal spray coating, AREGA.

[14] MAG'MAT, N° 22 / APRIL - JUNE 2007, technical dossier.

[15] Smith C.W., "The basic principles of flame spraying." Science and Technology of Surface Coating, 1974, pp 262-270.

[16] SOKOLOV D., "Contribution au développement de la projection thermique à très faible pression", Thesis, pp 11-12, 2009.

[17] Gausseem SAIF, PhD thesis, "mise au point et caractérisation de couches intermédiaires permettant l'accrochage entre un alliage métallique et une céramique projetée plasma", pp 8, 1991.

[18] Pfender E., (1992) "Fundamental Studies Associated with the Plasma Spray Process", Thermal Spray, Advances in Coatings Technology, C.C Berndt (Ed), (Pb.) ASM International, Materials Park, Ohio, USA, pp.1-10.

[19] Proner A., Revêtements par projection thermique, Techniques de l'ingénieur, Traité matériaux métalliques, M 1 645, pp 1-20.

[20] Fauchais P., Vardelle A. Dussoubs B., (2001) Quo Vadis Thermal Spraying Technology, Vol.10 (1), pp. 44-66.

[21] Korpiola K., Vuoristo P., (1996 "Effect of HVOF Gas Velocity and Fuel to Oxygen Ratio on the Wear Properties of Tungsten Carbide Coating", Practical Solutions for Engineering Problems, (Ed.) C.C. Berndt, (Pub.) ASM International, Materials Park, Ohio, USA, pp. 177-184.

[22] Arsenault B., Legoux J.G., Hawthorne H., (2001) "HVOF process optimization for the erosion resistance of WC-12Co and WC-10Co-4Cr Coatings", New Surfaces for a New Millennium, C.C Berndt Ed., Khor K. A., Lugscheider E., ASM International, Materials Park, Ohio, USA, pp. 1051-1060.

[23] GUILLAUME MAROT, Lille, PhD thesis, "Modélisation de l'essai d'indentation interraciale et confrontation aux essais normalisés pour la détermination de l'adhérence de revêtments obtenus par projection thermique", P 17, 2007.

[24] Bossard AG, Technical information: "Coating technology", pp: 3, 2009.

[25] ASTM D 907, Standard Terminology of Adhesives.

[26] J. CIZEK, K.A. KHOR, Z. PROCHAZKA, Influence of spraying conditions on thermal and velocity properties of plasma sprayed hydroxyapatite, Materials Science and Engineering C 27 (2007) 340-344

[27] L. PAWLOWSKI, The science and engineering of thermal spray coatings, Ed. Wiley, Chichester, UK, 1995.

[28] D.S. RICKERBY, A review of the methods for the measurements of coating-substrate adhesion, Surface and Coatings Technology 36 (1988) 541-557.

[29] PH. DÉMARÉCAUX, Adhérence et propriétés tribologiques de revêtments obtenus par projection thermique hypersonique : Applicabilité des revêtements de carbures de chrome aux disques de freins, PhD thesis, Université des sciences et technologies de Lille, 1995.

[30] D. H. James, "A review of experimental findings in surface preparation for thermal spraying", J. Mech Work Tech, Vol. 10, 1984, pp. 221-232.

[31] Y.-Y. Wang, C. -J. Li, A. Ohmori, "Influence of substrate roughness on the bonding mechanisms of high velocity oxy-fuel sprayed coatings", Thin Solid Films, Vol. 485, pp. 141-147, 2005.

[32] J.W. McBain, D.G. Hopkins, "Adhesion and Adhesives", J. Phys Chem, Vol. 29, pp. 188, 1987.

[33] L. Pawlowski, "The Science and Engineering of Thermal Spray Coatings", J. Wiley and Sons, 1995.

[34] D. Matejka, B. Benko, "Plasma spraying of metallic and ceramic materials", VCH Publishers, 1989.

[35] R. B. Heimann, "Plasma-spray coating: principles and applications", John Wiley & Sons, 1989.

[36] L. Pawlowski, "Optimisation des paramètres de projection des céramiques par plasma d'arc, Etude des propriétés physiques et thermiques des couches projetées, Exemple d'application : substrats pour la microélectronique hybride", Université de Limoges, Thèse d'état, 1985.

[37] A. Syed, P. Denoirjean, A. Denoirjean, J. C. Labbe, P. Fauchai, "Influence of substrate oxidation stage on the morphology and flattening of splats", J. Ther Spray, 2002.

[38] P. Fauchais, G. Montavon, M. Vardelle, J. Cerdelle, "Developments in direct current plasma spraying", Surf Coat Tech, Vol.201, pp. 1908-1921, 2006.

[39] R. Bonnet. Thermal spraying on complex shapes: simulation and calibration of the robotized process. PhD thesis. Université de Technologie de Belfort-Montbéliard, France, 2000.

[40] S. Guessasma. Optimization and control of thermal spray processes by cooperation of artificial intelligence methods. PhD thesis. Université de Technologie de Belfort-Montbéliard, France, 2003.

[41] F.I. Trifa. Deposition model for the simulation, design and realization of coatings developed by thermal spraying. PhD thesis. Université de Technologie de Belfort-Montbéliard, France, 2004.

[42] C.J. Li, B. Sun. Effects of spray parameters on the microstructure and property of Al2O3 coatings sprayed by a low power plasma torch with a novel hollow cathode. Thin Solid Films 450 (2004) 282-289.

[43] F.I. Trifa. G. Montavon, C. Coddet. On the relationships between the geometric processing parameters of APS and the Al2O3-TiO2 deposit shapes. Surface and Coatings Technology; Volume 195, Issue 1, 23 May 2005, 54-69.

[44] S. Guessasma, G. Montavon, C. Coddet. Modeling of the APS plasma spray process using artificial neural networks: basis, requirements and an example. Computational Materials Science, Volume 29, Issue 3, (2004) 315-333.

[45] JE Döring, R. Vassen, D. Stöver, The influence of spray parameters on particle properties, Proc. 2002 International Thermal Spray Conference and Exposition, E. Lugscheider, PA. Kammer (eds), DVS-Verlag GmbH, Düsseldorf, Germany, 2002, p. 440, 2002.

[46] A. Vardelle, Etude numérique des transferts de chaleur, de quantité de mouvement et de masse entre un plasma d'arc à la pression atmosphérique et des particules solides, Thèse de Doctorat, Université de Limoges, France, N° d'ordre : 29-87, 1987.

[47] J-F. Li, Modélisation de la formation des contraintes résiduelles dans les dépôts élaborés par projection thermique, PhD thesis. Université de Technologie de Belfort-Montbéliard, France, 2005.

[48] C.-J. Li, A. Ohmori, Relationship between the microstructure and properties of thermally sprayed deposits, Journal of Thermal Spray Technology, 11 (3), p.365, 2002.

[49] S-H. Leigh, Stereological investigation on structure/property relationships of plasma spray deposits, PhD Thesis, Department of Materials Science and Engineering, State University of New York, Stony Brook, USA, 1996.

[50] M. Prystay, P. Gougeon, C. Moreau, Structure of plasma sprayed zircona coatings tailored by controlling the temperature and velocity of the sprayed particles, Journal of Thermal Spray Technology, 10 [1], p. 67, 2001.

[51] K.A. Khor, Y.W. Gu ; D. Pan, P. Cheang. Microstructure and mechanical properties of plasma sprayed HA/YSZ/Ti-6Al-4V composite coatings. Biomaterials 25 (2004) 4009-4017.

[52] R. Bonnet, O. Landemarre, R. Bolot, C. Coddet, Thermal spray coating simulation - step 2, Report No. 01026-020830, LERMPS, France.

[53] J. Ilavsky, A. Allen, G.G. Long, S. Krueger, Influence of spray angle on the pore and crack microstructure of plasma-sprayed deposits, Journal of the American Ceramic Society, 80, p. 733, 1997.

[54] M.F. Smith, R.A. Neister, R.C. Dykhuizen, An investigation of the effects of droplet impact angle in thermal spray deposition, Thermal Spray Industrial Applications, C.C. Berndt, S. Sampath (eds.), ASM International, Materials Park, OH., USA, p. 603, 1994.

[55] G. Montavon, S. Sampath, C.C. Berndt, H. Herman, C. Coddet, Effects of the spray angle on splat morphology during thermal spraying, Surface and Coatings Technology, 91, p. 107, 1997.

[56] J. Ilavsky, A. Allen, G.G. Long, S. Krueger, Influence of spray angle on the pore and crack microstructure of plasma-sprayed deposits, Journal of the American Ceramic Society, 80, p. 733, 1997.

[57] V.V. Sobolev, J.M. Guilemany, Flattening of droplets and formation of splats in thermal spraying: a review of recent work- Part 2, Journal of Thermal Spray Technology,8(2), p. 301, 1999.

APPENDIX

APPENDIX A :

I. Attenuation:

An ultrasonic wave loses energy as it propagates through a real medium. Intrinsic attenuation is caused by one of two phenomena: absorption (dissipation of energy in the form of heat) and diffusion (porosities, inclusions, grain boundaries, defects, etc.). The principle of ultrasonic attenuation measurement by contact reflection is based on the acquisition of several successive echoes. The piezoelectric transducer generates a beam of plane waves which reflect off the bottom of the workpiece and return to the transducer. The attenuation then shows an exponential decrease in amplitude of the form $A_{2max} = A_{1max} \exp(-\alpha x)$ With x=2*e.

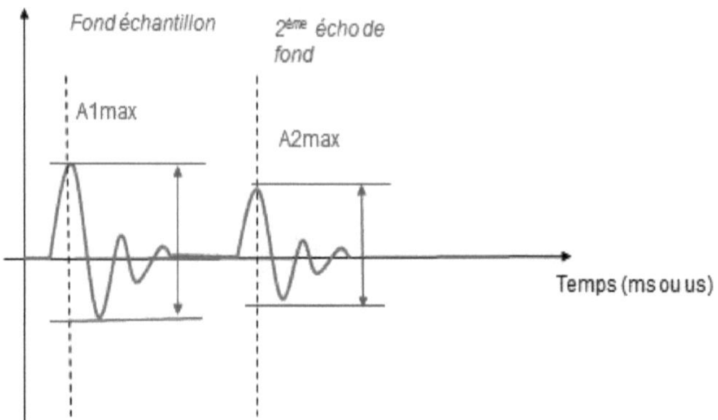

$$So \quad \alpha = \frac{1}{2e} \ln(\frac{A_1}{A_2})$$

II. Description of ECHORGAPHE 1085 and ECOM'85 software:

1. Description of ECHORGAPHE 1085.

ECHOGRAPH 1085, shown in Figure US-C1, consists mainly of :

- ➢ a display for viewing the various parameters and the received signal.
- ➢ an **"On/Off"** button to switch the unit on or off.
- ➢ two C1, C2 connectors to link the device to the translator(s):

✓ the C1 connector is used for echo control;

✓ C2 connector is used for transparent control. The transmitter is connected to C1 and the receiver to C2.

➢ two **"Range"** buttons:

✓ Knob R1 adjusts the horizontal offset (i.e. the delay assigned to the displayed signal);

✓ The R2 button for scaling (i.e. setting the size of the display window for the received signal).

➢ Three **"Gates"** buttons:

✓ The G1 button moves the cursor horizontally, allowing you to read the position, in time, of the echoes of a given signal;

✓ The G2 button changes the size of the cursor, allowing you to read the time between two echoes of the same signal;

✓ The G3 button for vertical displacement, allowing you to read the amplitude of echoes.

➢ A **"menu"** button to view and, if necessary, change device parameters.

➢ Two **"Gain"** buttons:

✓ The Ga1 button for amplitude amplification ;

✓ The Ga2 button for setting rejection. Rejection is a threshold that can be applied at will to display only amplitudes above this threshold. Its purpose is to eliminate background noise (grass). Its use is not recommended, as it can eliminate echoes relating to small defects.

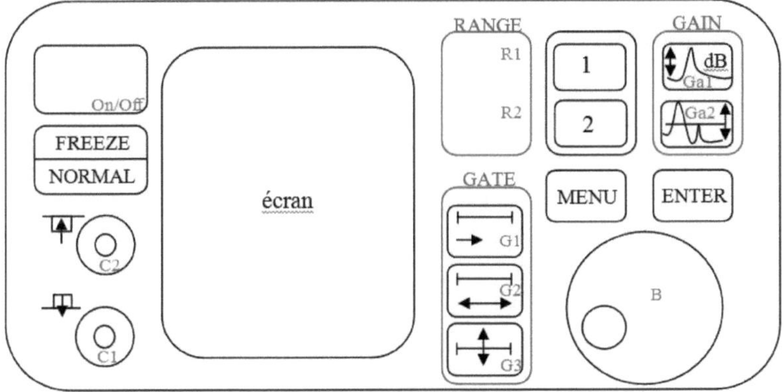

Figure: Schematic description of the 1085 ultrasound scanner.

To change the value of a given parameter, simply click on the corresponding button, then turn knob B in either direction to increment or decrement its value.

Example: to increase signal amplitude, click on the Gal button, then turn knob B to the right.

2. ECOM'85 software.

The ECHOGRAPH 1085 is connected to a PC to exchange data and record the signals received by the sensor. This transfer is carried out using a software program called "ECOM'85". The main panel of this software is shown in figure US_I1. It comprises 5 buttons:

- ✓ **Configuration"** button for setting data transfer parameters ;
- ✓ **Report"** button for data transfer and signal display ;
- ✓ **Parameters"** button to view the various control parameters;
- ✓ **Text"** button to write a report on the acquisition;
- ✓ **Data"** button for transferring control parameters.

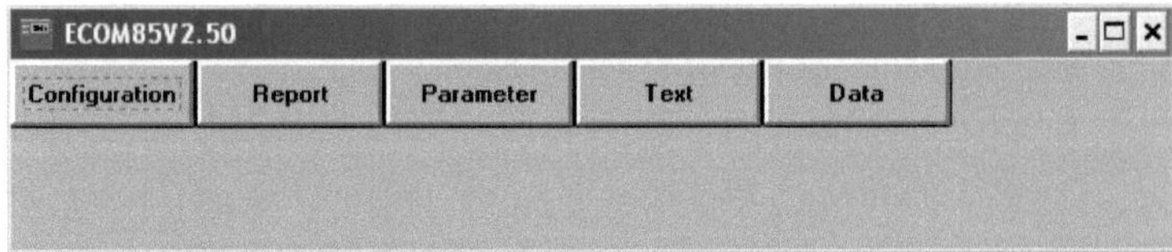

Figure US_1: ECOM'85 software main panel

This software transfers signals from the ECHOGRAPH 1085 to the PC. To do this, two steps are required:

➤ **1ère step: click on** the **"Configuration"** button, the window shown in figure US_2 appears. This window allows you to set the various transfer parameters, as well as the desired language:
- ✓ The port connected to the Echograph: select COM1;
- ✓ Data transfer speed: select 9600 baud ;
- ✓ Language: choose French ;
- ✓ Ultrasound scanner type: choose Echograph 1085.

Once you have done this, click on the "Save" button. The "OK" button lights up green.

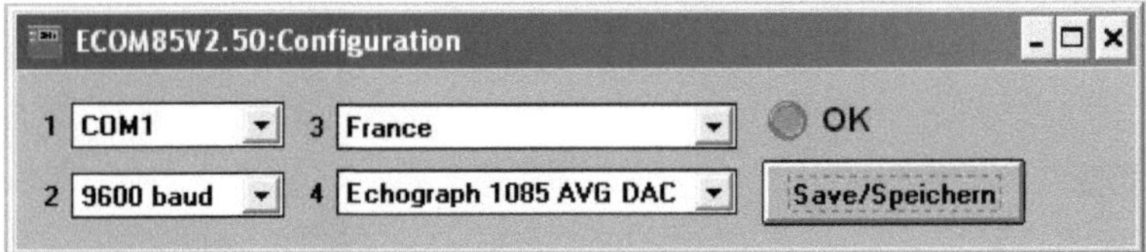

Figure US_2: Configuration window

> ➤ **2ème** step: click on the "**Report**" button in the main panel, the window shown in figure US_3 appears. In this window, click on the "**Transfer << 1085**" button.
If the "Radio" button lights up red, there's a problem; if it lights up green, the transfer has been completed correctly. Click on the "A_SCAN" button to view the transferred signal (figure US_4).

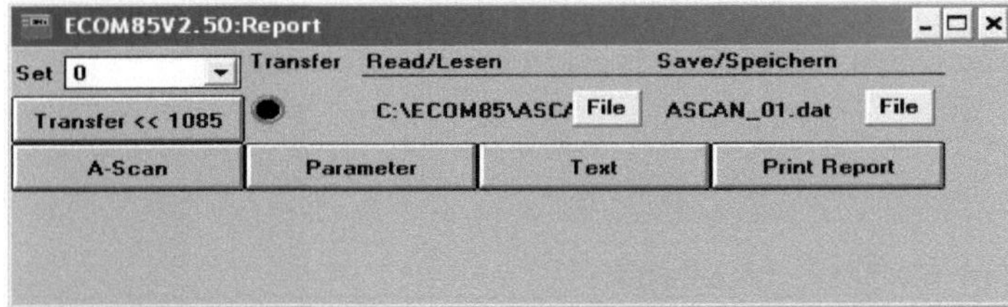

Figure US_3: "Report" window

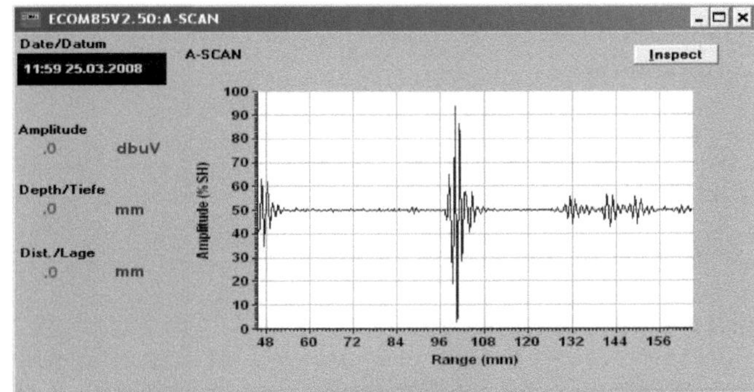

Figure US_4: "ASCAN" window

APPENDIX B :

I.   Comparison of attenuation in different materials of the same thickness for two transducers (5MHz and 2MHz) :

a. Steel :

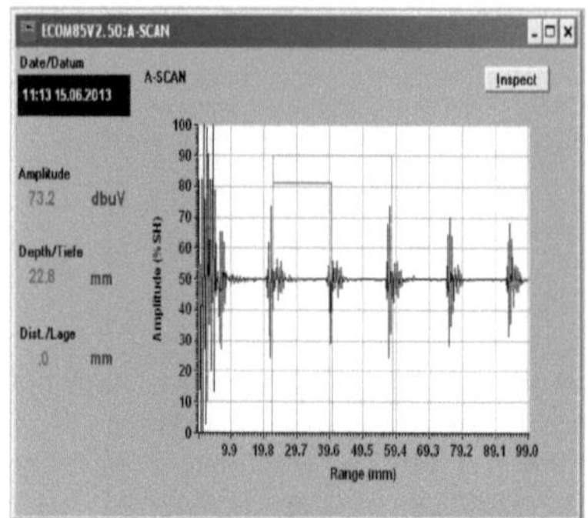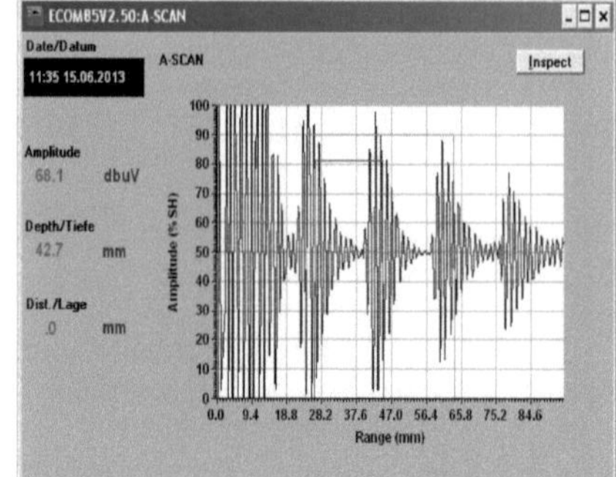

Steel 5MHZ Steel 2MHZ

The attenuation calculation is given by the following formula:

$$\alpha = \frac{1}{2e}\ln(\frac{A_1}{A_2}) \qquad\qquad (1)$$

For the 5Mhz translator :

Part thickness e = 18mm

To determine the amplitude, we take two successive echoes, then differentiate between the maximum and minimum.

Amplitude:

$A_1 = (74\text{-}23.5) = 50.6db$

$A_2 = (70\text{-}23.5) = 46.6db$

En remplace dans la formule (1), on obtient :

$$\alpha = 0.00234mm^{-1}$$

For the 2Mhz translator :

$A_2 = 48db$ and $A_1 = 60\ db$

by replacing in formula (1), we obtain :

$$\alpha = 0.00619 mm^{-1}$$

b. Aluminium :

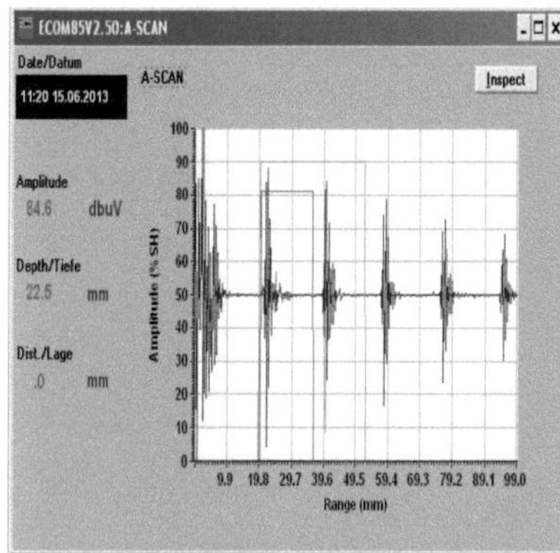 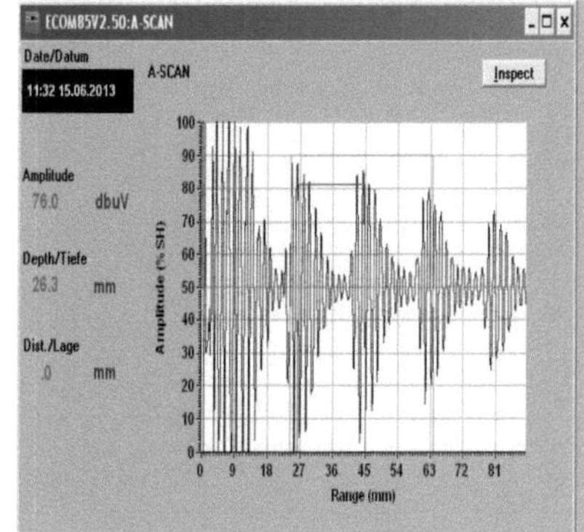

Aluminium 5MHZ Aluminium 2MHZ

c. Plastic :

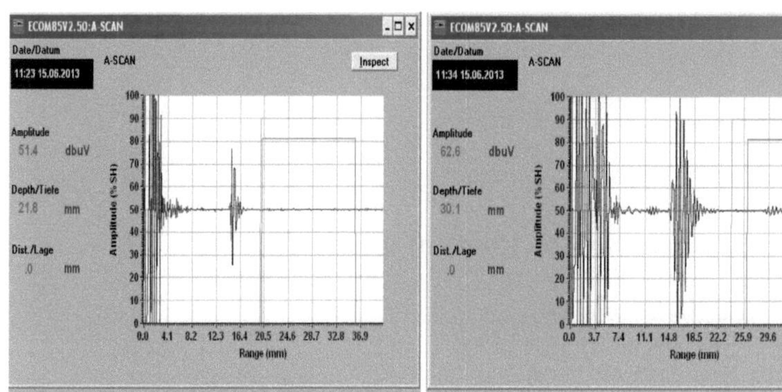

Plastic 5MHZ Plastic 2MHZ

d. Bronze :

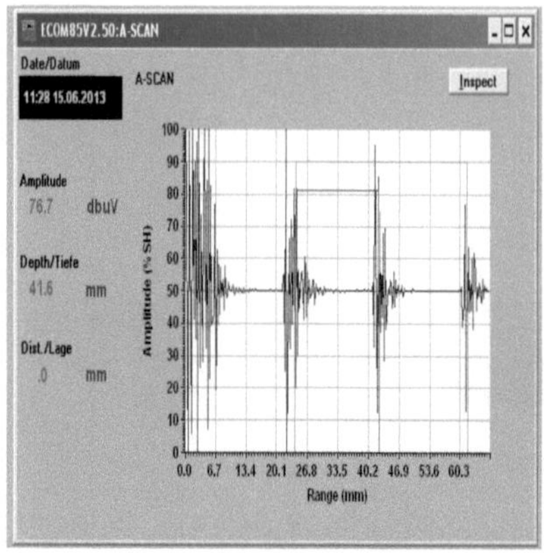
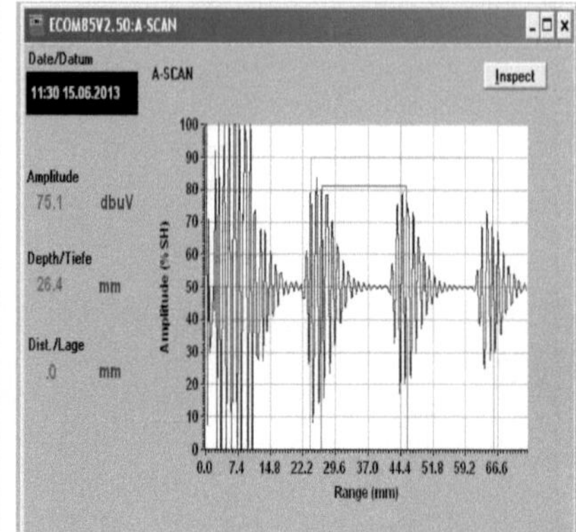

Bronze 5MHZ Bronze 2MHZ

	Steel	Aluminum	Bronze	Plastic
$\alpha(5Mhz)en(mm^{-1})$	0.0023	0.0015	0.0178	——
$\alpha(2Mhz)en(mm^{-1})$	0.006	0.00506	0.0066	——

Metals / *Attenuation*

II. Calculates the thickness of a defect.

> **Translator parameters**
> - ✓ The thickness is 24 *mm*
> - ✓ The type of translator: focus
> - ✓ Frequency: 2 *MHz*
> **Sample parameters**
> - ✓ Thickness
> - ✓ Propagation speed

***Sample*: Aluminium**

Sample thickness: 65.5 *mm*

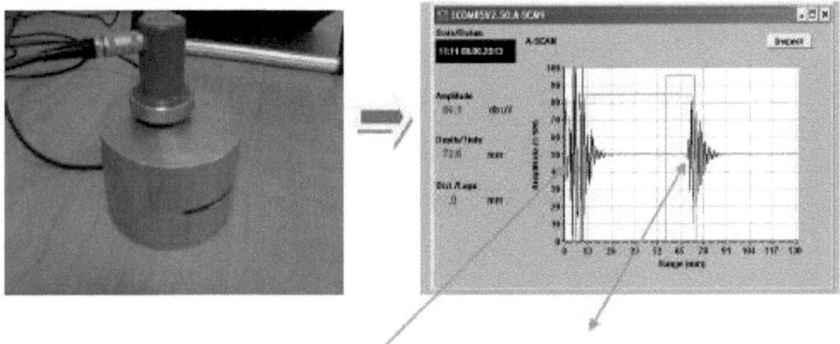

Echo1 de la réflexion de surface   **Echo2** de la réflexion de fond

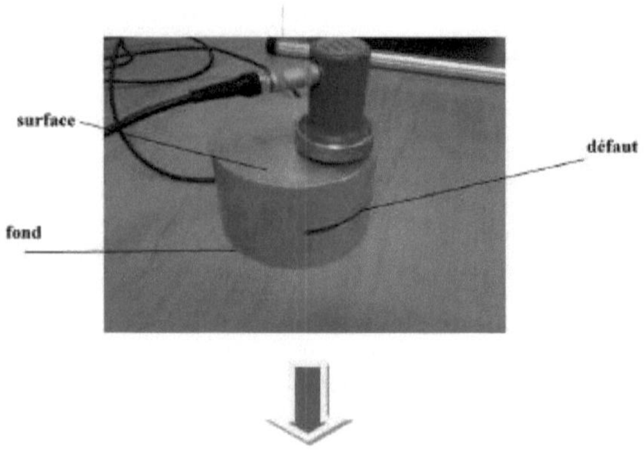

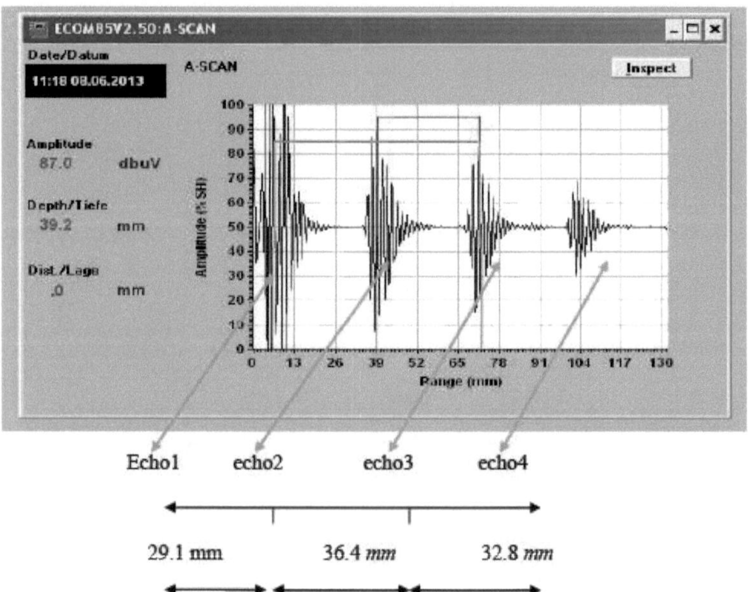

Defect thickness = 65.5 - (29.1 + 32.8) = 3.6 *mm*

Echo1 surface reflection.

Default Echo2.

Echo3 for in-depth reflection.

Echo4 fault (the same fault).

Conclusion

Measuring the position of defects by ultrasonic inspection is not a simple matter, and measuring defect volume is very complex, but image processing can be used to detect defect contours.

yes
I want morebooks!

Buy your books fast and straightforward online - at one of world's fastest growing online book stores! Environmentally sound due to Print-on-Demand technologies.

Buy your books online at
www.morebooks.shop

Kaufen Sie Ihre Bücher schnell und unkompliziert online – auf einer der am schnellsten wachsenden Buchhandelsplattformen weltweit! Dank Print-On-Demand umwelt- und ressourcenschonend produzi ert.

Bücher schneller online kaufen
www.morebooks.shop

info@omniscriptum.com
www.omniscriptum.com

Printed by Books on Demand GmbH, Norderstedt / Germany